DIGITAL TRANSFORMATION KEEPING IT SIMPLE

DIGITAL TRANSFORMATION , INDUSTRY 4.0 , ANALYTICS AND THE LATEST IN CUTTING EDGE ENERGY SOLUTIONS

SAMEER PIKALE

Made with ♥ on the Notion Press Platform
www.notionpress.com

I would like to dedicate this technology related book , " Digital Transformation Keeping it Simple" to Girish Mehendale technocrat, founder and the force behind Microverse Automation India's only Indigenous DCS manufacturer. I also thank Lalit Shaiwale one of my best friends and my professional mentor. Siddharth Mehendale for supporting our common Mission Madness to create an Indian Global Leader in Automation and Digitization. Thanks to everyone at Microverse family we surely will achieve this dream.

#prodlymadeinindia

#aatmaniirbharbharat

#indiantechnology

Contents

Contents

Foreword

They say behind every successful man is a woman. In my case there are 4. The first one is my mother Priyamvada Pikle who gave birth to me and inculcated good values in me. The 2nd is my daughter Netra Pikle who taught me to love competitive sports like Taekwondo. She is a fighter and taught me to never give in any adverse situations in life. The third is my little one our little princess Bhakti Pikle, who with her infectious smile and bubbly nature taught me to spread happiness and love unconditionally. The last but not the least is my wife my soulmate Kashmira Pikle. She is my pillar of strength my anchor in turbulent times. She has always stood behind me as rock of support. Always ready to walk besides me and egging me on to achieve greater success. Thank you all for being my family my inspiration.

I would also like to thank my elder brother Milind Pikle whom I always admire for his calm demeanour and his perfection. My extended family, my in laws Nitin Desai (Pappa) for his great advices and solid backing. Namrata Desai (Mummy) for her all-pervasive love. Rachna (sis in law) and Nazim (Saadu Bhai) my most de-stressing and jolly couple. Siddhant (bro in law) and Tanmeyi (little sis in law) for dotting on my kids.

Thank you to everyone at my Rotary Club, Rotary Club Of Ponda New Generation Goa for taking me in the Rotary Parivar and giving me a purpose in life. Making me believe in Service above self and serving the community to change lives.

Thank you all.

A small appeal to buy this book and allow me to donate to the most transparent and successful philanthropic organization The Rotary Foundation (TRF). The Rotary Foundation is a non-profit corporation that supports the efforts of Rotary International to achieve world understanding and peace through international humanitarian, educational, and cultural exchange programs. It is supported solely by voluntary contributions.

For every 1 USD earned from the sale of this book I will donate 0.5 USD to the TRF and also match 0.25 USD from myside to the TRF. So, guys and gals make this happen. An appeal to give back to the society we live in a very small way.

#digitaltransformation
#technology
#feelinginspired
#communityservice
#proudlyindian
#rotaryclubofpondanewgeneration

Preface

Hi everyone this is me, Sameer Pikale, an Engineer by profession. A hobby artist by nature, dabbling in water colour paintings in between busy work schedules and a writer by heart, living his dreams penning passionate blogs and technology related articles.

Professionally I have always been Passionate about Digital Technologies, IOT, Analytics, Big Data, Cloud, SaaS & PaaS and over all Technology related solutions 25 years of extensive experience in Industrial Automation, Manufacturing & Infrastructure IT Solutions Sales, Business Development, Advance Analytics Software and Asset Performance Management Software & Services Sales.

Well versed with Consultative high-value Business outcome selling to C-Level Executives. Expertise is in Driving New Business growth in the Process Industry (O&G, Chemical, Power, Metal) with a PAN India exposure handling Direct Sales & Channel Partners. Proven Track record in achieving and exceeding Order & Sales Plan along with Contribution Margin responsibilities.

Proficiency is in Industrial Internet of Things (IIoT), Manufacturing Operation Management, Advance Analytics, Asset Performance Management, SCADA, and Control System solutions.

Expertise is in organizing Techno-commercial Resources to carry out business operations and have a clear understanding of developing new businesses in Indian & Middle East.

Hands-on experience is in exploring new markets, identifying the right customer's and partner to accelerate business growth in India & Middle East Market.

Exhibited the ability to lead the teams of Solution Architect, Proposal/Estimation, and Engineering professionals to achieve set Sales & Marketing targets.

In-depth knowledge of operations in various Process Industries like Power, Chemical, Metal, Cement, O&G & Life Science. Competent Communicator and Negotiator having strong analytical, problem solving & organizational abilities.

Travelled for business to Dubai and America.

About this book and why I chose to write " Digital Transformation- Keeping it Simple “. This book is a collection of my articles on Digital Transformation and cutting-edge technologies, my professional experiences and trying to make the technical jargons simple to understand and easy to read. A lot of has been said by many experts on technologies like Artificial Intelligence, Predictive Maintenance, Reliability Centred Maintenance, Virtual Reality etc. I have tried to break it down and explain using simple used cases. Same goes for the new age energy sources and digitization techniques. Hope you like this book.

Happy Reading...

CHAPTER ONE

DIGITAL TRANSFORMATION .... KEEPING IT SIMPLE.

Open any technical journal or a magazine or even on social media like LinkedIn the most trending topic is all about Digital Transformation, Industry 4.0, Industrial Internet , IoT , Machine Learning , Cloud Computing , AI or AR and how they are helping the industries and to achieve their Business Outcomes. All these jargons mean only one thing being connected and using data to take informed decisions. Decisions that transform the way we run our businesses.

To understand all this we need to know what is Industry 4.0 is all about First revolution came when Steam engines were invented and steam was used to power the first machines that mechanized some of the work our ancestors did. Next revolution was when electricity was discovered, this was used to power up the assembly lines which gave birth to mass production. The third revolution of industry came about with the advent of computers and the beginnings of automation, when robots and machines began to replace human workers on those same assembly lines.

And now we enter Industry 4.0, in which computers and automation are integrating together in an entirely new way building an ecosystem with Robotics, Artificial Intelligence (AI) and Augmented Reality (AR) along with machine learning algorithms that can self-learn and control the smart devices and machinery to achieve business outcome with very little or no manual interventions. All this is possible over remote networks and getting connected to computer systems on the plant shop floor.

Industry 4.0 introduces what has been called the "smart factory," in which cyber-physical systems monitor the physical processes of the factory and make decentralized decisions. The physical systems become **Internet of Things**, communicating and cooperating both with each other and with humans in real time via the wireless web.

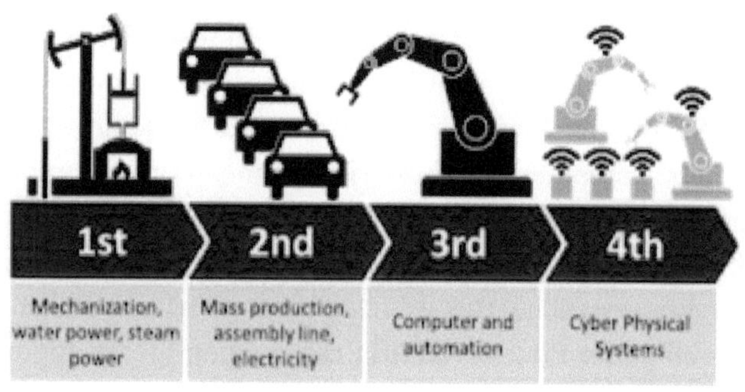

Industrial Revolution 4.0

For a factory or system to be considered Industry 4.0 ready, it must include:

· **Big Data** – Huge amounts of data from disparate islands of information, machines , manufacturing systems , ERP etc.

· **Interoperability** — machines, devices, sensors and people that connect and communicate with one another.

· **Information transparency** — the systems create a virtual copy or what we call Digital Twin of the physical/ real world through sensor data in order to Contextualize information.

Technical assistance — both the ability of the systems to support humans in making decisions and solving problems and the ability to assist humans with tasks that are too difficult or unsafe for humans.

· **Decentralized decision-making** — the ability of Digital systems to make simple decisions on their own and become as autonomous as possible.

Simply put to roll out Industry 4.0 or a Digital Road Map for any organization is a 3 step process to be implemented in phases. These are **Capture, Contextualize and Control**. Each phase is important in its own respect and any organization will be in any of states of readiness. Some ahead of others with Integrated Plants and some in very nascent stage of getting their plant floors connected for data capture. To elaborate these 3 phases.

Capture – This is the very basic and 1st stage where data has to captured from all know sources of information be it from machines, manual operated work stations, Quality labs systems, Utilities, Supply chain and Business Systems. The data coming from these systems is huge may be in Tera Bytes and are generated at different velocity and in different varieties. Conforming to the **3 V's ofBig Data –**

Volume , Velocity and Variety

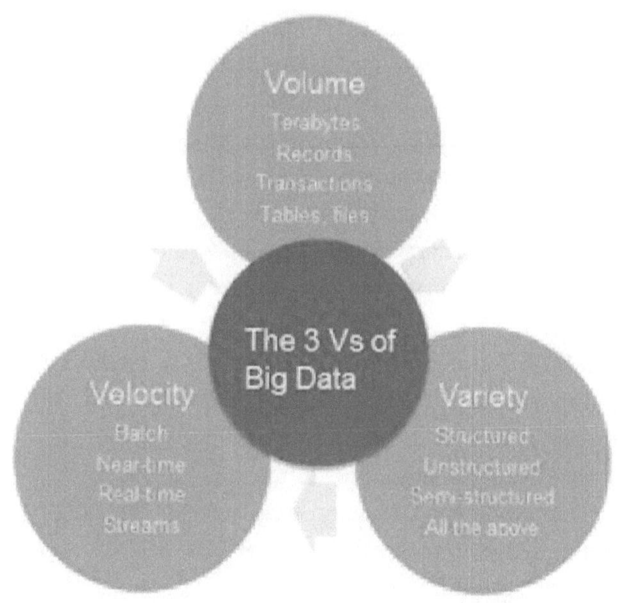

3 V's of Big Data

The data being Industrial in nature majority of it is related to **Process e.g. Temperature, pressure, Quality e.g. Volume, density, viscosity** and some in the form of transactional data which is manually entered. All these needs to stored in a different way than the traditional Relational databases. To effectively store and recall you will need a database which can store for long periods of time (years together) and in huge volumes and yet be able to be retrieved at a click of a mouse without latency. This database can be on premise or now with advance technology in the Cloud. The cloud offers the user to reduce his overall cost of digitization by offer him the chance to pay per use. Also the system administration and IT setup cost is taken out of the equation.

The 2nd phase is the important phase, that of **Contextualize.** Contextualize is to get meaningful insights from all the data being captured. This gives the person looking at the data an additional perspective to help in quick decision making. One level above the standard data that is visualized in form of the traditional SCADA screens or Plant Dash Boards. Insights catering to reasons of line stoppages, reasons of unplanned downtime, reduced efficiencies of workforce to effectively and proactively plan the production schedules. Also, insights on which products to be run on which lines to maximize productivity. Here the key is to get both Real time information for as is Information as well as Historical data to analyse best times and best operating condition so as to be operational consistency which I turn will lead to increased production and reduce cost of operation.

The 3rd and last stage is **Control or what we call the Optimization stage.** After analysing the data both real time and historical you can understand the gaps in the current processes and find areas of improvements. It could be as simple as **Digitization of SoP's** for smoother task management and improve the workforce efficiency. Or insights of Energy Consumption leading to implementation projects for better utilization of utilities in terms of use steam and water. Or decision on whether to switch between cheap Grid power to costly Captive Power using diesel or furnace oil as fuel. Understand reasons of bad quality during material processing and identifying golden parameter to achieve consistency to get best quality products at all times.

This today is being achieve by Machine learning algorithms which mine the Big Data to come out with process rules to help in Optimization either automatically by way of an Advance Process Control or Semi-automatic decision support systems or Intelligent Alarming.

Another example is to use machine learning for assets management by way of Predictive and Prescriptive Analytics what we know as **APM or Asset Performance Management**. Take the instance of a tube leakage or a exceptional bearing temperature rise in a critical component of a Turbine. If this alert is given to a Combined Cycle Power plant Operations & Maintenance team say about 30-40 days before it is going to happen it will lead a huge savings 1stt in terms of no unplanned downtime and 2nd in terms of preventing unnecessary opening up of capital equipment's leading to increased time of shut downs and 3rdreduces the cost of huge spares inventories. This in turn reduces the cost of generation and increases uptime which in turn affects the bottom line of the Power generation business.

But as with any major shift, there are **challenges** inherent in adopting an Industry 4.0 model:

1. Data security issues are greatly increased by integrating new systems and more access to those systems. Additionally, proprietary production knowledge becomes an IT security problem as well.

·2. A high degree of reliability and stability are needed for successful interfacing between plant equipment and the Cloud as inter communication can be difficult to achieve and maintain.

·3. Maintaining the integrity of the production process with less human oversight could become a barrier.

4. Loss of high-paying human jobs is always a concern when new automations are introduced.

·5. And avoiding technical problems that could cause expensive production outages is always a concern.

Additionally, there is a lack of experience and manpower to create and implement such systems — not to mention a general reluctance from stakeholders and investors to invest heavily in newer technologies.

But the benefits of an Industry 4.0 model could outweigh the concerns for many production facilities. In very dangerous working environments, the health and safety of human workers could be improved dramatically. Supply chains could be more readily controlled when there is data at every level of the manufacturing and delivery process. Computer control could produce much more reliable and consistent productivity and output. And the results for many businesses could be increased revenues, market share, and profits.

Reports have even suggested that emerging markets like India could benefit tremendously from Industry 4.0 practices, and the city of Cincinnati, Ohio has declared itself an "Industry 4.0 demonstration city" to encourage investment and innovation in the manufacturing sector there.

The question, then, is not if Industry 4.0 is coming, but howquickly. As with big data and other business trends, I suspect that the early adopters will be rewarded for their courage jumping into this new technology, and those who avoid change risk becoming obsolete and redundant.

So to put this into simple perspective **Industry 4.0 is nothing but smart way to Capture Contextualize and Control** data / processes leading to reduces cost and improving efficiency and achieving Business Outcomes.

As they say **Being Simple** is so **Difficult** but then **Simple** is **always Beautiful...**

CHAPTER TWO

Reliability Centred Maintenance (RCM) the future of Proactive Maintenance Management

In the last couple of years there has been a clamour for Prescriptive and Predictive Maintenance to move away from traditional Time-Based Maintenance to Condition Based Maintenance especially in Asset heavy Industry Verticals of Oil & Gas, Petrochemicals, Power and Metal Mines Minerals. For some time now Industry experts have been advocating use of Data Driven technology combined with Advance Analytics to gain insights into Asset behaviour its diagnostics vide data that gets captured in Process Systems and other systems like Vibration Monitoring Systems and use it for decision making.

Typically, analytics is used for smart notifications and alerts to give a window of opportunity to the Operations and Maintenance Team to pre-empt anomalies in advance in any asset condition before it turns into a catastrophe. But how you consume these Early Warning Notifications / Alerts sets the tone for Proactive Maintenance Management.

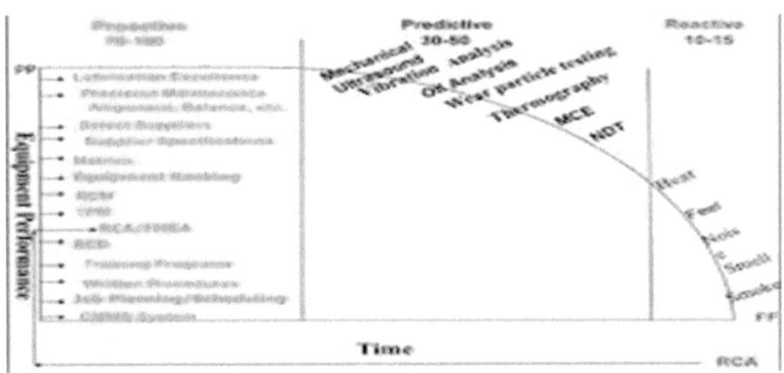

Proactive Predictive Curve

This is where Reliability Centred Maintenance (RCM) comes into picture. **Reliability Cantered Maintenance (RCM)** is a concept of maintenance planning to ensure that systems continue to do what their user require in their present operating context. Successful implementation of RCM will lead to increase in cost effectiveness, reliability, machine uptime, and a greater understanding of the level of risk that the organization is managing. Ultimately, by performing RCM, organizations are looking to develop unique maintenance schedules for each critical asset within a facility or organization.

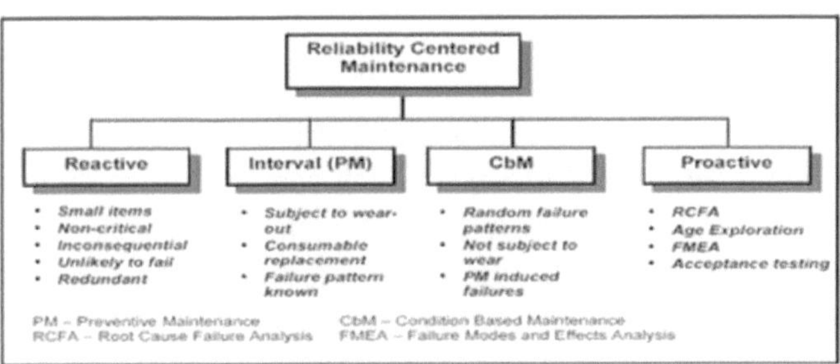

RCM Stages

There are four basic principles of an RCM program, stated in different ways by organizations all over the world. A program is RCM if it:

1) Is designed, scoped and structured to preserve system function
2) Identifies failure modes, which are the ways in which something might fail. Failures are any errors or defects, especially ones that affect the customer, and can be potential or actual
3) Addresses failure modes by importance
4) Defines applicable maintenance task candidates and selects the most effective one in the case of important failure modes

Industry Experts / Domain Consultants have described an RCM program as:

- A process that "uses a cross-functional team to develop a complete maintenance strategy designed to ensure inherent design reliability for a process or piece of equipment."
- A way "to identify components whose functional failures can cause unwanted consequences to one's plant or facility."

As per the technical standard SAE JA1011, Evaluation Criteria for RCM Processes, which sets out the minimum criteria that any process should meet before it can be called RCM. This starts with the seven questions below, worked through in the order that they are listed:

1. What is the item supposed to do and its associated performance standards?

2. In what ways can it fail to provide the required functions?

3. What are the events that cause each failure?

4. What happens when each failure occurs?

5. In what way does each failure matter?

6. What systematic task can be performed proactively to prevent, or to diminish to a satisfactory degree, the consequences of the failure?

7. What must be done if a suitable preventive task cannot be found?

Reliability cantered maintenance is an engineering framework that enables the definition of a complete maintenance regimen. It regards maintenance as the means to maintain the functions a user may require of machinery in a defined operating context. As a discipline it enables machinery stakeholders to monitor, assess, predict and generally understand the working of their physical assets. This is embodied in the initial part of the RCM process which is to identify the operating context of the machinery, and write a **Failure Mode Effects and Criticality Analysis (FMECA).** The second part of the analysis is to apply the "RCM logic", which helps determine the appropriate maintenance tasks for the identified failure modes in the FMECA. Once the logic is complete for all elements in the FMECA, the resulting list of maintenance is "packaged", so that the periodicities of the tasks are rationalised to be called up in work packages; it is important not to destroy the applicability of maintenance in this phase. Lastly, RCM is kept live throughout the "in-service" life of machinery, where the effectiveness of the maintenance is kept under constant review and adjusted in light of the experience gained.

RCM can be used to create a cost-effective **maintenance strategy** to address dominant causes of equipment failure. It is a systematic approach to defining a routine maintenance program composed of cost-effective tasks that preserve important functions.

The important functions (of a piece of equipment) to preserve with routine maintenance are identified, their dominant failure modes and causes determined and the consequences of failure ascertained. Levels of criticality are assigned to the consequences of failure. Some functions are not critical and **are left to "run to failure"** while other functions **must be preserved at all cost**. Maintenance tasks are selected that address the dominant failure causes. This process directly addresses maintenance preventable failures. Failures caused by unlikely events, non-predictable acts of nature, etc. will usually receive no action provided their risk (combination of severity and frequency) is trivial (or at least tolerable). When the risk of such failures is very high, RCM encourages (and sometimes mandates) the user to consider changing something which will reduce the risk to a tolerable level.

The result is a maintenance program that focuses scarce economic resources on those items that would cause the most disruption if they were to fail.

RCM emphasizes the use of **predictive maintenance(PdM)** techniques in addition to traditional preventive measures.

How Do You Implement A Reliability Cantered Maintenance Program?

There are three phases (Decision, Analysis and Act) of a reliability cantered maintenance program , and seven steps within these phases to ensure the program is fully implemented.

Phase I: Decision

Justification and planning based on need, readiness and desired outcomes.

1. Analysis Preparation

RCM analysis is only as effective as the team behind it. The most effective cross-functional teams include maintenance employees, project leaders, subject matter experts, and if possible, executive leadership.

Additionally, documenting procedures and your project plan can be vital to keeping your team on track. The beginning of an RCM project is a great time to outline your organizational goals, project management concerns, budget and timeline, and potential obstacles.

2. Select Equipment for RCM Analysis

Equipment selected for RCM analysis should be critical to operations, the cost of repair vs. replace and previous spending on Preventive Maintenance. To select the best candidate, ask yourself these questions:

- Could failure be difficult to detect during normal operation and maintenance?
- Could failure affect safety?
- Could failure have a significant impact on operations?
- Could failure have a significant impact on spending?

3. Identify Functionality

Define a complete list of a piece of equipment's functionality, including as much data driven information as possible. It is important for your team to specify your desired asset performance levels instead of actual performance, as it may reflect an operational or maintenance issue. System functionality then drives the required functions of the equipment supporting the system functions.

Phase II: Analysis

Conduct the RCM study in a way that provides a high quality output.

4. Identify Functional Failures

Functional failure is the inability of an asset or system to meet acceptable standards of performance. Failures can encompass poor performance, over performance, performing unnecessary or unintended functions, or complete failure.

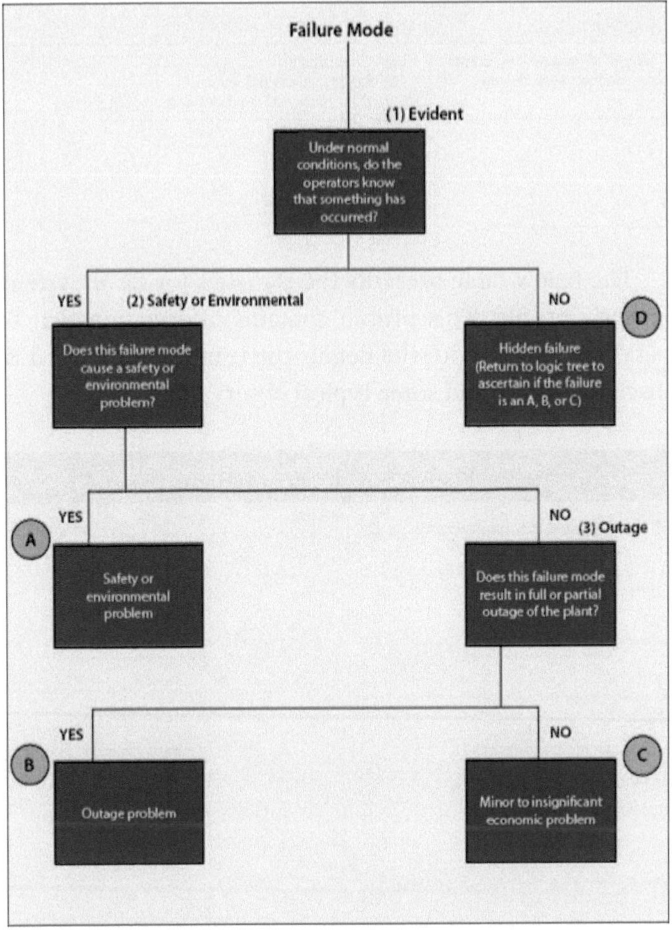

Failure Modes

For example, when a motor bearing is failing due to lack of lubrication, a Total Functional Failure would be the motor not rotating, and the motor failing to function.

Function	Functional Failure
Deliver oil from tank to tank at no less than 500L/minute	1) Oil is leaking 2) No oil is being delivered 3) Oil is being delivered slower than 500L/minute

Function Failure

The below table presents the statistics for RCM System analysis profile. This profile contains information that is very descriptive with the details the team has examined & discussed. Here and some typical observations

Table 2 – Statistics from RCM System Analysis Profile		
RCM Systems Analysis Profile	**Centrifuge Subsystem**	**Waste Heat Boiler Subsystems**
Subsystem Functions	6	7
Subsystem Functional Failures	9	11
Components in Subsystem Boundary	16	25
Failure Modes Analyzed Critical Non-Critical Hidden	46 29 (63%) 17 (37%) 13 (28%)	63 28 (44%) 35 (56%) 7 (11%)
PM Tasks Specified (includes Run to Failure)	62	70
Active PM Tasks	53	58
Items of Interest	30	16

Statistics from RCM Analysis

5. Identify and Evaluate the Effects of Failures

Next, your team should document what actually happens when failures occur. What can be observed? What is the impact of the failure on production? Is there a significant safety impact?

6. Identify Failure Modes

Once you identify your equipment and systematic functional failures, failure modes must be considered. One of the most common techniques to approach discovering failure modes is Failure mode and effects analysis (FMEA). FMEA is a step-by-step approach for identifying all possible failures in a design, a manufacturing or assembly process, or a product or service. Understanding the effects of failure involves asking questions such as:

- What are the safety concerns with this failure?
- What impact does this failure have on operation/production?
- Does this failure mode result in full or partial outages?

A CMMS offers automation tools to help reduce missing scheduled work and equipment failures, making PM optimization as efficient and streamlined as possible. PM Task Generation, PM Scheduling and Inspections help facilitate continuous improvement and support for an organization's Preventive Maintenance program.

Phase III: Act

Act on the study's recommendations to update asset and maintenance systems, procedures and design improvements.

Design Improvements in RCM

7. Select Maintenance Tasks

At this point, the most appropriate maintenance action can be identified based on the failure mode information. Failure management techniques can be grouped into two categories:

- **Proactive tasks** – Preventive and Predictive Maintenance techniques are performed to prevent failure of a piece of equipment of system. Preventive Maintenance is calendar or usage based, and helps to reduce the risk of failure, while Predictive Maintenance, or Condition Monitoring, can detect the failure before it begins.
- **Default actions** – Fire fighting or reactive maintenance deal with failures after the fact. Run-to-failure is a tactic where equipment is run until it fails, and then work is

performed.

Selecting the right strategy for failure management is rooted in an understanding of failure modes, criticality of equipment and the economic impact of failure.

Economic Impact of Failure

RCM implementation with a CMMS

A properly implemented Computerized Maintenance Management System (CMMS) can support the RCM process. CMMS helps maintenance programs develop goals for cost tracking, benchmark data and monitor the bottom line. For example, with proper Online reporting and dashboard tools, organizations can consistently document work order history, failures, costs and trends. With a few clicks of a mouse, organizations have access to the data to perform the analyses that RCM requires.

The goals of RCM include the ability to evaluate, categorize, prioritize and understand the appropriate way to intervene in the impact of failures. Ultimately, by performing RCM, an organization is looking to develop unique maintenance schedules for each critical asset within a facility or organization.

Successful implementation of a RCM process, coupled with CMMS software, will increase cost effectiveness, asset reliability, equipment uptime, and an enhanced understanding of the level of risk that the organization is managing.

If you go by popular sentiments then if Preventive Maintenance and Predictive maintenance is the need of the hour in achieving Operational Excellence but Reliability Cantered Maintenance (RCM) is the future of Proactive Maintenance Management.

CHAPTER THREE

SOP Digitization As a 1st Step towards Digital Transformation

Standard Operatig Procedure

Engaging and Supporting the Operator of Today into the Future with Guided Process Control.

As businesses try hard to gain a competitive edge against their business competitors in today's complex market scenarios the way operators perform their work in their manufacturing operations, has evolved drastically. Shifting from being purely functional to highly analytical, and becoming smart operators has become a necessity and an integral part of today's business processes—significantly changing the way businesses operate and add value to archive their business outcomes.

Operators earlier were required to be proficient in following set procedures and achieve maximum production. Current operators are expected to be cross functional and go beyond duty to achieve stated business outcomes. In short be agile.

To think Digital Transformation is all about data and data driven technology would be belittling the contribution of the shop floor workers in this journey towards achieving Operational Excellence. For any improvement in overall efficiency of a plant or OEE any solution so proposed should take into consideration in the ease of doing work on the manufacturing lines / shop floor with minimal instructions and maximum compliance to Standard Operating Procedures (SoP)

To advance operator effectiveness, the latest Process Automation Controllers (PAC) enables operators to better leverage information in both routine and critical conditions for optimal decision making. Such intelligent PAC's satisfies the growing need for tools that enable them to collect, connect, analyse, and act upon vast amounts of real-time operations data, supporting a more engaged and intelligent workforce.

History of Modular Procedural Automation

A procedural operation consists of a **set of operator tasks that are conducted in a set way time-after-time to achieve a certain goal such as starting or shutting down a unit or making a product**. As demonstrated by Paul McKenzie (Bristol Myers Squibb) at the World Batch Forum (WBF) North America conference 2007 in Baltimore, it could even apply to the operation of an analyser. So how was the journey to modular procedural automation?

If you look back at the functions available to control systems for both programmable logic controllers (PLC) and distributed control systems (DCS), they focused on discrete and continuous control. Sequential and batch control functions were added later and the logic was more complex. **Batch processes are more procedural in nature**, but typically involve sets of procedures running in parallel on varying process units and in multiple process stages which almost always need to have built in flexibility. The **ISA-88 standard** ultimately addressed batch automation very well; however, attempts were made in the 1970s and 1980s to implement advanced procedure management using control system functions with strings of function blocks or by straight-line coded applications. These functions certainly could handle the sequential nature of procedure management but they had limited flexibility.

Current Industry Challenges:

One key area is in alerts and communicating to the operator. Work Process management and control require a complex interface but control systems by their very nature limit this ability. Therefore, providing limited information to the operator hampered his ability to make procedure decisions, take action on alternatives, and realize the reasons for current process states and reasons for process deviations. Managing a procedure or process transition in automatic mode was non-effective and operators chose to accomplish procedure management manually.

Another difficult area to overcome was that **most Standard Operating procedure (SOP) management involves unsteady state operation** which includes many manual operations for the field and plant floor operator. This meant that the skill and knowledge of running the SOP stayed with the operators and operations staff and not with the control system.

The Chemical industry with **both continuous and batch processes** began implementing SOP digitization. Procedure and process transition flow charts with steps guided the operator from operator consoles.

Transition Management

In its basic definition, transition management is a **method of implementing advanced procedure management.** The goal is to improve overall performance through faster and smoother transitions. **This contributes to extended equipment life and optimized production and increased product yields.** There are three broad operational procedures involving transitions: **start up, shutdown and unit state change.** The unit state change includes transitions like grade changes, production rate changes, process equipment switches, etc. Transition management applies equally to a complete process unit and to a selected piece of equipment in a processing unit, like a shutdown valve, or a heat exchanger. Transitions also inherently increase the risk of disruptions that can lead to incidents or lost production. Incidents that occurred during start up and shutdown continued to be a major factor. An additional study showed that examination of major incidents by the average loss per incident indicated that operational error caused the largest average dollar loss.

A fourth area that is added to transition management is **"abnormal" process state changes**, like a process trip resulting in a asset shutdown or a larger event causing a production shutdown. In some abnormal transitions, like a process trip, most of the time there is a window of opportunity for a process operator to reset the equipment and bring the operating conditions to normal. But this window is narrow needs to be effectively used else it results in longer shutdown and higher production losses.

Capturing & Creating Procedural Knowledge Base

The existing paper-based operator procedures are "static" in nature. The actual procedural knowledge and skill is with the operators. However, **knowledge and the accompanying skill sets** to run these procedures manually is diminishing. How to preserve the knowledge of the best operators and their practices is a challenge. **SoP Digitization** can help in capturing their activities and create a knowledge base. Level of technological complexity is increasing, the timely flow of information, data, and knowledge is more important than ever in the process industries.

Procedural Operations

The chemical facilities are run and maintained according to operational procedures and every process a SOP. A SoP has a set of tasks that are conducted in a sequence and in set time. These procedures require consistent execution. There are three main types of procedures: manual prompted and automated.

1. Manual

Manual procedures are performed by operators taking actions either in the control room or in the field. Often operators perform these procedures based upon their training and experience. There is always a variability associated with manual operations. Documentation is necessary for process improvement, compliance and additional data analysis.

2. Guided

Guided procedures are implemented in a control or manufacturing execution system which steps through the procedures, stopping at the end of each step or action and waiting for the operator to manually grant permission to continue. Prompted procedures can decrease variability, transition times and enable automatic record keeping.

3. Automated

Automated procedures are implemented in control systems and generally only stop at the end of normal operational sequences. If an abnormal situation takes place or an operator intervenes, the sequence can be held or perform exception handling. Automated procedures reduce variation and time.

Conclusion

The work already achieved in chemical industries has proven the value of SoP digitization. It creates operational knowledge base, process consistency which reduces operator dependence and improves production.

CHAPTER FOUR

AI –AR THE FUTURE OF MAINTENANCE IN MANUFACTURING ARENA

Nearly a decade ago 2002 to be precise a cyber action thriller by the name of "Minority Report" starring Hollywood's superstar Tom Cruise and directed by the Sci-Fi master Steven Spielberg was released and went on to become that years blockbuster movie. The theme was very unique and unusual. Set primarily in Washington, D.C., and Northern Virginia in the year 2054, where Precrime, a specialized police department, apprehends criminals based on foreknowledge (predictive analysis) provided by three psychics called "precogs". The cast includes Tom Cruise as Chief of Precrime John Anderton, Colin Farrell as Department of Justice agent Danny Witwer, Samantha Morton as the senior precog Agatha, and Max von Sydow as Anderton's superior Lamar Burgess.

AI - AR

The film combines elements of tech noir, whodunit, thriller and science fiction genres, as well as a traditional chase film, as the main protagonist is accused of a crime he has not committed and becomes a fugitive. Spielberg has characterized the story as "fifty percent character and fifty percent very complicated storytelling with layers and layers of murder mystery and plot". The film's central theme is the question of free will versus determinism. It examines whether free will can exist if the future is set and known in advance. It also shows the role of preventive government in protecting its citizenry, the role of media in a future state where technological advancements make its presence nearly boundless, the potential legality of an infallible prosecutor.

There are many scenes in which Tom Cruise uses the capabilities of the Precogs to analyse the data captured by the Precogs in the form of their visions and analyses them in Virtual Reality on screens made up of Digital Images in thin Air. Much like the present-day VR & Augmented Reality device projections.

Look around you and all this has come true with the present day IoT, Machine Learning and AI – AR technologies. Companies today have more access to data than ever before in history. This is expected to increase as the cost of the core connectivity devices become cheaper and more accessible worldwide. From accounting to HR, employee training programs to product manufacturing and delivery, businesses are looking to continue leveraging the data at their disposal to make their internal process better, faster, and more efficient.

As expected, they are already using these technologies to address one of the most challenging aspects of running an organization that owns physical assets - efficient maintenance management. Of particular interest is the fact that these technologies allow communication from machine-to-machine (M2M) and from machine-to-humans (M2H). However, the fact that machines are now able to think intelligently, learn, teach themselves, make decisions and respond like (or even better) than humans, has launched a never-before-seen opportunity to improve recurrent maintenance issues.

Today, there is profound interest in the ways to use technologies like Artificial Intelligence (AI), Machine Learning (ML), Virtual Reality (VR) or Augmented Reality (AR), and the Industrial Internet of Things (IIoT) to reduce costs and improve asset optimization and workers' safety.

The following is a look at how the face of maintenance is changing with the introduction of these technologies.

Artificial Intelligence in Maintenance Management

The cost of unplanned equipment downtime is staggering. Take the manufacturing sector for instance - the International Society of Automation (ISA) estimates that manufacturers lose $647 billion globally every year to downtime.

To combat this challenge, companies are now taking a more proactive approach to their maintenance strategy. Instead of waiting for equipment to fail before repair (**reactive maintenance**) or replacing parts on a strict time-based schedule (**preventive maintenance**), they are using intelligent devices and systems to predict and address problems before they occur (**predictive maintenance**).

Over the years, the shortcomings of time-based maintenance have become obvious. A Boeing study shows that up to 85 percent of equipment fail despite regular calendar-based maintenance. Predictive maintenance came as a result of the need to improve on these shortcomings.

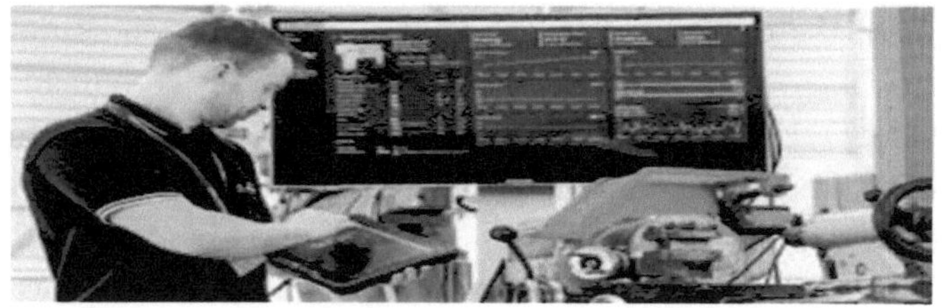

AI based maintenance

Artificial Learning and Predictive Maintenance: How It Works

AI software:

- Captures data from sensors, exiting systems (DCS / PLC) , third party softwares like QCS , Vibration Analysis Systems etc
- Monitors anomalies and patterns (historical benchmarks combined with data analytics to give signature patterns for both assets and process)
- Uses that data to request for human intervention by way of Early Warning Notifications and Alerts

For predictive maintenance, there are two ways AI monitors the information coming out of assets:

Anomaly: The system reads data generated from the equipment and picks up any variations from normal running conditions.

Failure: This focuses on data patterns to detect a potential failure based on similarities with predefined failure modes.

Both of these generate alerts to let the maintenance team know that there's a need to investigate the situation. When combined, they form the basis of advance maintenance strategies likecondition-based maintenance and predictive maintenance.

Predictive maintenance is now being used to monitor a wide variety of equipment and it has the major advantage of picking up on faults that a human inspector would miss, or simply wouldn't be able to inspect or test. Better yet, companies in manufacturing, mining, oil and gas, and public utilities are using this technology to remotely monitor their critical assets scattered all over the world on land and underwater. This saves them from the inconvenience of sending their staff out for routine inspections as was the case in past years.

Other Maintenance Applications for AI

Inventory Management - Automated inventory management means organizations can minimize, or eliminate, the need for manual inventory checking. This removes the common problem of human error and time wasted while checking large stocks of items.

AI also allows the tracking of any product or spare part in real-time. But it shouldn't necessarily end with tracking of inventory items alone. Take a look at Ocado, a British online-only supermarket where over a thousand robots work all day to fill grocery shopping orders from buyers. Facilities could be developed in the near future where spare parts and other maintenance inventory items are sorted and issued to technicians entirely by robots.

Fleet Management - Artificial intelligence has become a game-changer for fleet management. The growing adoption of AI means fleet and maintenance managers can now achieve more control over the most pressing challenges they face especially with regards to fuel management, vehicle maintenance, logistics, unauthorized use of vehicles, safety and driver performance and tracking.

The possibilities are still being developed but it's already obvious that this technology delivers better-managed fleets, considerable cost savings, and improved productivity.

Augmented Reality in Maintenance

Companies are using virtual reality / augmented reality technology in their maintenance efforts majorly for employee training. At the forefront of this trend are manufacturers and major aviation giants includingAir France, Airbus, and Boeing.

It's easy to understand why. For example, the manufacturers want to empower their workforce and plant-operators to maintainfactories of the future (FoF). But, the assets in smart factories are typically extremely expensive and sensitive.

How do they train new workers to operate machinery without jeopardizing the integrity of the equipment or endangering the individual? Simple. Virtual reality and augmented reality devices are solving this problem.

Instead of hiring experts or designating more experienced staff to train new entrants every time, these businesses are creating unique training experiences that are accessible just by wearing a VR headset or operating AR apps on tablets and smartphones. The trainee is able to interact with a customized simulated environment and experience a first-person view.

AR based maintenance

An added advantage is that these training are accessible on-demand basis 24/7 and for multinationals, their staff can train themselves from any of their facilities anywhere around the globe. Or even from the comfort of their homes in their free time.

In terms of Asset Maintenance the field force can now carry a Smart Tablet which is pre-configured with the software solution, uses Advanced analytics of machine and fleet data to Deliver role specific KPIs, alarms, and trends to the mobile user.

These new age software solutions provide Mobile KPIs, analytics, and collaboration for better service delivery. Structured navigation and views Making the most of mobile KPI presentation. Navigation driven by asset model and user role. Makes setup for each machine simple – creates templates by machine type. Combined machine state and performance Offers mobile KPI data in a single, accessible view. On a single screen it provides

1. Key asset nameplate information
2. Key relationships
3. Health index indicators
4. Alarm state aggregation
5. Key performance indicators
6. Real time charting

Such solutions Improve fleet management using predictive alerts and with geo-intelligence – proximity awareness its Drives O&M teams' responsiveness

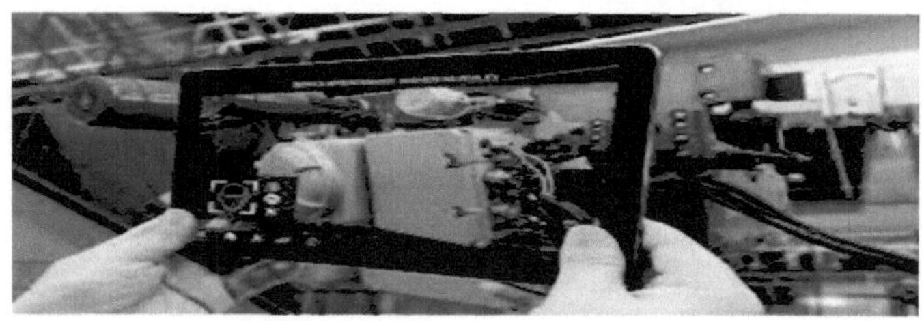

AI -AR based apps for maintenance

What is the Future?

Going into the future, the implications of AI for maintenance in whatever industry will likely be in the form of more automation and responsibilities for the machines rather than humans.

The thought and decision-making process of machines, especially robots, will continue to advance with several benefits such as carrying out work in dangerous or hazardous environments, the use of self-driving vehicles for better supply chain management (SCM), and monitoring humans to prevent fatigue-related accidents.

No one knows exactly what futures going to look like but definitely, maintenance in tandem with artificial intelligence AI and Augmented Reality AR will be better and more efficient than what was available in the past.

CHAPTER FIVE

WFH – Work From Home – The new age Digital Office spaces

WFH Work from home ..that's the one word that has been most used these days. And rightly so what with the entire world grappling with the global pandemic due to Covid 19.

Work From Home

But how do business's work effectively in these tough times. Earlier offices were needed for word collaboration between working teams. For team leaders to discuss and delegate work and to have meetings with clients to understand project requirements and let them know updated status of running projects.

Sales teams needed the office for preparing commercial offers, sending out marketing mails, fixing appointments and follow-up with clients and prospects alike. But things started changing with remote site support and client morning meetings especially in the Software & Services Industry. With clients also approving remote support and even project implementation and executions. Remote Technical support centres became the new way of professional life.

But still the need for a dedicated office space for in house departments like HR, Finance, and SCM / Procurement to name a few was always there. These were last in the line to adopt technological enhancements. A few exceptions would be the roll out of Enterprise ERP systems like SAP / Oracle which was implemented more for its financial visibility & MIS reporting a much-needed tool for the top management for decision making.

Most organizations are yet to put their Digital Strategy in place. In such a situation technical leader within the organizations take decisions in bits and pieces. Sometimes even on hearsay and other people's views. They then start using different applications to cater to different functional requirements. Typical requirements would be for a Video App for team / client meetings. A mailing app for sending and receiving mails and replying to queries. Another application would be for SCM & Logistics for tracking Inventory and Finished goods in the Warehouse as also physical movement of goods in transit from factory warehouse to distributor depots. Finance department will require a separate application for Billing & invoicing and payroll management while HR would require an application for employee on boarding and training. This is your first response that is to find some quick technology fixes. You quickly adopt a messaging and video conferencing platform to replace your current conversations. Then you will keep approving more tools that come your way that seem to facilitate better collaboration.

But you are actually in the middle of a highly condensed version of what CIOs of digital companies have been struggling with for the last 10 years. Transitioning to a digital culture doesn't happen overnight and not by adopting a few tools. It requires a complete strategy overhaul. The obvious solution isn't always the right one. If you are new to digital work, your 1st instinct is to pursue a new tool for each problem you are trying to solve. You follow a best of the breed model and use tools that are well known for certain functions. However, this quickly causes an overload of applications and creates even more confusion. Employees will spend their time jumping from one application to the next and constantly learning new interfaces. You will need t expand your IT team just to handle all the chaos and to build custom connections between different tools. In the end adding more digital solutions often results in increased costs not only in software but also in the number of people who need to manage it.

With the additional telecom bandwidth availability and improvement in remote connectivity the scenario has changes. In the current situation there is no option but to put all your organizational functions on Remote Operations. But it is easier said than done.

The strategy for every organization small or big should be to have an Enterprise Digital Transformation template which has to be modular, scalable and flexible to the changing processes and requirements.

At the high level these Digital solutions should emanate from a single Digital platform with modular functionality that can be enabled as you move ahead in your business life cycle. And be such that it empowers you personnel rather than hinder their day-to-day operations.

Yes, there will be reluctance but everyone has to fall in line as this is the new normal.

Some of the business challenges that every organization will face if a common platform is not implemented are –

1. No clear enterprise visibility
2. Lack of integration / co-ordination
3. Longer time for decision making
4. Reduces outputs
5. Reduced profitability.

Take for example a service support team using **Zoom** (the new darling app) for video meetings. A **Microsoft outlook** for mailing and **Jira or Trello** for task coordination. Though each in its individual capacity serves its purpose in reality it is increasing data creation in silos with no inter-reference. Tomorrow if a client complaint of slow or no response to its list of punch points the service team has to go through realms of mail and service reports to establish any sense of trail for support given.

Imagine if all this is available in a single Digital platform. Life would be much simpler and more efficient. But as technology improves all this is possible take for example the **Google Cloud** as your digital platform. It has a mail feature under **Gmail**, a team collaboration tool like **Hangout**, a video application **Duo** for video conferencing and the best part is all this can be analysed with the best-in-class analytics tool – the **Google analytics**. Same is the case if you a **Microsoft** fan, you have the **Azure** as cloud platform, **Office 365** for office functions like documentation – Word, spreadsheets – Excel, presentation tools – PowerPoint and many more. Collaborative tools like

Microsoft Teams, Skypefor business for meetings and video conferencing.

So, all that is needed is a will to side-line the instinct to take the easy way and implement disjointed applications. Also, in the long run cost of maintaining multiple & diverse applications is much higher than the expected cost benefits.

As some wise men have already said "Change is the only this that is constant in Life". The more you accept this scenario easier it will be to survive and also surpass.

Here's to WFH the New Normal....

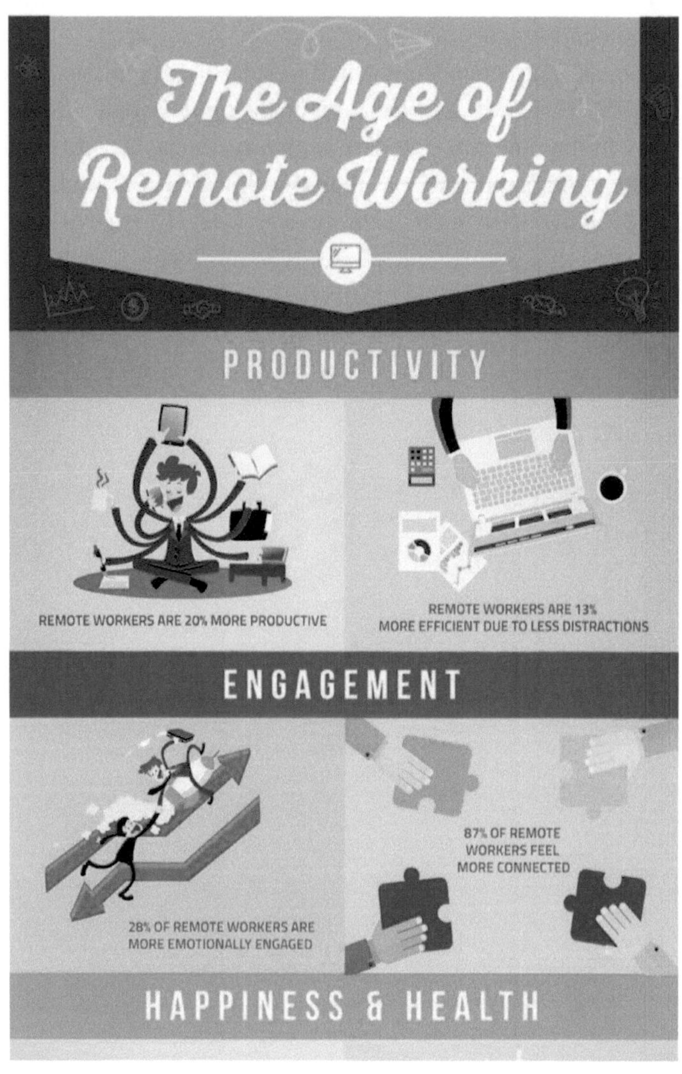

The Age of Remote Working

CHAPTER SIX

VIRTUAL ASSISTANT'S ... THE FUTURE IS CALLING

Imagine this you wake up in the morning, you know at the back of the mind the days going to be tightly schedules, you are a little stressed out you call out to your voice activated Wi-Fi connected speaker to play your favourite Retro FM station. The speaker takes your command switches to x **107.2 FM Radio Nasha FM Radio** station which belts out old Dev Anand Classics .... **Pal Bhar Ke Liye Koi Hume Pyaar Karle Jhoota hi Sahi** ... A smile erupts on your lips. You go to the kitchen and call out to the Coffee Machine to brew you a strong cuppa. You sip on the piping hot Coffee and ask your Jeeves your man Friday on your iPhone to relay you the Top 10 headline news of the day. You check out the news then the time and rush to take a shower before you head out to office. While getting dressed you again check Google Maps for the traffic details to decide whether to take your car or book a aggregator cab, then you remember you have to fly to another city in the late evening. No point taking your car. You ask Jeeves to book you an Uber to the office. Just as you are finished dressing and have picked up your overnighter bag with you most precious cargo, your Netbook, you get a message prompt on your mobile that your Cab has arrived. You softly tip toe into your bedroom, kiss your Sweet Wife and your little princess good bye and head for the door. You take the lift down, alight the cab and you are on the way to your office an hour's drive from where you live.

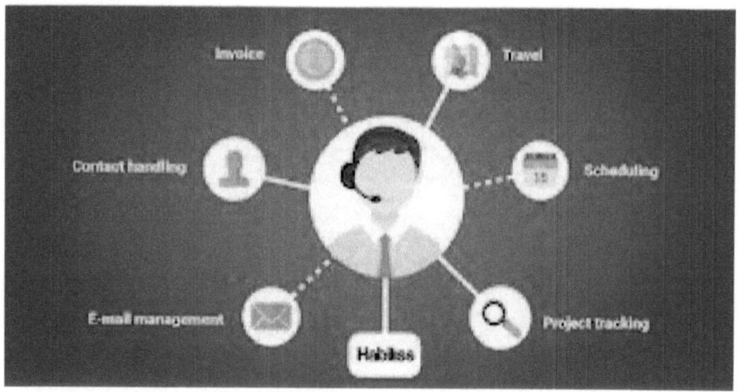

Virtual Assistant

You again ask **Jeeves** to check you mails for important mail ids and sort them on priority. You spend the next half hour replying to these mails. Suddenly you remember you have to prepare a business report for your boss taking inputs from your colleagues and subordinates. Again, Jeeves is there to rescue. You dictate a mail to Jeeves reminding 3 of your colleagues and 2 of your subordinates asking them for individual business reports to be sent to you by 11 AM so you can cull the final report and mail it to the Boss.

You reach the office and exchange some pleasantries with the ever-smiling receptionist before getting behind your workstation. You have a schedule meeting in one of your conference rooms with you Distributor in half an hour. You again ask **Jeeves** to co-ordinate with the Facilities Management System to prepare the conference room with proper heating control, LCD projector, Tea and some Biscuits. All by way of mails from your account. A prompt on your iPhone Calendar indicates

the impending birth day of your wife and since you are away for the next two days on a trade show you have to get her some terrific gift. You call out to Jeeves ask him to search your favourite Web Market Place "Amazon" for a Handbag from Caprese or a Habbo from Larvie. Jeeves checks your order history for the last hand bags bought for your wife, checks the latest stock available, the best possible deals and suggests 2 options for you a Fuchsia Pink Hobo or a Black Party Hand bag. You select the later and place the order.

Its 10:30 AM and you go in to the conference room to meet personnel from your distributor. Shake hands and few head nods and its business as usual. Half way through the meeting you excuse yourself for 15 minutes check your mails, combine the individual reports you received into a single report, check it cross check it and then mail it to the Boss. The lasts for 2 hours and you end it on a spirited note and optimism. You shake hands and escort your guest to the reception. By now you are famished and need to have a quick bite. You again take help of Jeeves to order a Subway and a Salad. You chit chat with you colleagues in the neighbouring work stations before your food order arrives. You grab a quick bite and rush for the next meeting at a client using the same aggregator Cab and Jeeves to help you off course. During the meeting you get an alert on your iPhone about the fight delay due to holiday congestion. You ask Jeeves to check out your Company Flight & Hotel booking app to try and reschedule your flight for an alternate time. You get a better deal on a good time; you make the transaction for new booking and cancellation and also dictate a mail to your hotel at the destination regarding your rescheduled time so they can send a pick up accordingly. You wind up the meeting and head to the Airport sighing in relief of the day getting over and looking forward to a bright day next morning....

So what do you think this is a scene from a Hollywood Sci-Fi movie or Reality... Welcome to 2017 the digital age an age of integrated world...Voice activated Virtual assistant's using Artificial Intelligence.

Voice Commands

The Jeeves I was talking about is real, its **SIRI** s an intelligent personal assistant, part of Apple Inc.'s **iOS, WatchOS macOS and tvOS operating systems**. The assistant uses voice queries and a natural language user interface to attempt to answer questions, make recommendations, and perform actions by delegating requests to a set of Internet services. The software adapts to users' individual language usages, searches, and preferences, with continuing use. Returned results are individualized.

Siri is a spinoff from a project originally developed by the pin-SRI International Artificial Intelligence Centre. Its speech recognition engine is provided by Nuance Communications, and Siri uses advanced machine learning technologies to function. Its original American, British, and Australian voice actors recorded their respective voices around 2005, unaware of the recordings' eventual usage in Siri. The voice assistant was released as an app for iOS in February 2010, and it was acquired by Apple two months later. Siri was then integrated into iPhone 4S at its release in October 2011. At that time, the separate app was also removed from the iOS App Store. Siri has become an integral part of Apple's products, having been adapted into other hardware devices over the years, including newer iPhone models, as well as iPad, iPod Touch, Mac, Air pods, Apple TV and the upcoming Home Pod.

Siri supports a wide range of user commands, including performing phone actions, checking basic information, scheduling events and reminders, handling device settings, searching the Internet, navigating areas, finding information on entertainment, and is able to engage with iOS-integrated apps. With the release of iOS10 in 2016, Apple opened up limited third-party access to Siri, including third-party messaging apps, as well as payments, ride-sharing, and Internet calling apps. With the release of iOS 11, Apple has updated Siri's voices for more clear, human voices, supports follow-up questions and language translation, and additional third-party actions.

Siri's original release on iPhone 4S in 2011 received mixed reviews. It received praise for its voice recognition and contextual knowledge of user information, including calendar appointments, but was criticized for requiring stiff user commands and having a lack of flexibility. It was also criticized for lacking information on certain nearby places, and for its inability to understand certain English accents. In 2016 and 2017, a number of media reports have indicated that Siri is lacking in innovation, particularly against new competing voice assistants from other technology companies. The reports concerned Siri's limited set of features, "bad" voice recognition, and undeveloped service integrations as causing trouble for Apple in the field of artificial intelligence and cloud-based services; the basis for the complaints reportedly due to stifled development, as caused by Apple's prioritization of user privacy and executive power struggles within the company.

The other famous voice activated app is **Alexa** is an intelligent personal assistant developed by Amazon, first used in the Amazon Echo and the Amazon Echo Dot devices developed by Amazon Lab126. It is capable of voice interaction, music playback, making to-do lists, setting alarms, streaming podcasts, playing audiobooks, and providing weather, traffic, and other real-time information, such as news. Alexa can also control several smart devices using itself as a home automation system.

Most devices with Alexa allow users to activate the device using a wake-word (such as *Echo*); other devices (such as the Amazon app on iOS or Android) require the user to push a button to activate Alexa's listening mode.

In November 2014, Amazon announced Alexa alongside the Echo. **Alexa was inspired by the computer voice and conversational system on board the Starship *Enterprise* in science fiction TV series and movies, beginning with *Star Trek: The Original Series* and *Star Trek: The Next Generation.***

The name Alexa was chosen due to the fact that it has a hard consonant with the X and therefore could be recognized with higher precision. The name is also claimed to be reminiscent of the Library of Alexandria, which is also used by Amazon Alexa Internet for the same reason. In June 2015, Amazon announced Alexa Fund, a program that would invest in companies making voice control skills and technologies. The US$100 million in funds has invested in companies including Ecobee, Orange Chef, Scout Alarm, Garageio, Toymail, MARA, and Mojio. In 2016 the Alexa Prize was announced to advance the technology.

In January 2017, the first Alexa Conference took place in Nashville, Tennessee, an independent gathering of the worldwide community of Alexa developers and enthusiasts. The follow-up has been announced, to be keynoted by original Amazon Alexa / Connected Home product head Ahmed Bouzid.

At the Amazon Web Services Reinvent conference in Las Vegas, Amazon announced Alexa for Business and the ability for app developers to have paid add-ons to their skills.

Alexa offers weather reports provided by AccuWeather and news provided by TuneIn from a variety of sources including local radio stations, NPR, and ESPN Additionally, Alexa-supported devices stream music from the owner's Amazon Music accounts and have built-in support for Pandora and Spotify accounts. Alexa can play music from streaming services such as Apple Music and Google Play Music from a phone or tablet. Alexa can manage voice-controlled alarms, timers, and shopping and to-do lists, and can access Wikipedia articles. Alexa devices will respond to questions about items in the user's Google Calendar. As of November 2016, the Alexa Appstore had over 5,000 functions ("skills") available for users to download, up from 1,000 functions in June 2016.As of a partnership with fellow technology company, Microsoft, Alexa will be available via its competing virtual personal assistant, Cortana. This functionality will come later in 2017.

But the best one is the Google Assistant...**Google Assistant** is a virtual personal assistant developed by Google and announced at its developer conference in May 2016. Unlike Google Now, the Google Assistant can engage in two-way conversations.

Assistant initially debuted as part of Google's messaging app Allo, and its voice-activated speaker Google Home. After a period of exclusivity on the Pixel and Pixel XL smartphones, it began to be deployed on other Android devices in February 2017, including third-party smartphones and Android Wear, and was released as a standalone app on the iOS operating system in May. Alongside the announcement of a software development kit in April 2017, the Assistant has been, and is being, further extended to support a large variety of devices, including cars and smart home appliances. The functionality of the Assistant can also be enhanced by third-party developers.

Users primarily interact with the Google Assistant through natural voice, though keyboard input is also supported. In the same nature and manner as Google Now, the Assistant is able to search the Internet, schedule events and alarms, adjust hardware settings on the user's device, and show information from the user's Google account. Google has also announced that the Assistant will be able to identify objects and gather visual information through the device's camera, and support purchasing products and sending money, as well as identifying songs.

The Google Assistant, in the nature and manner of Google Now, can search the Internet, schedule events and alarms, adjust hardware settings on the user's device, and show information from the user's Google account. Unlike Google Now, however, the Assistant can engage in a two-way conversation, using Google's natural language processing algorithm. Search results are presented in a card format that users can tap to open the page. In February 2017, Google announced that users of Google Home would be able to shop entirely by voice for products through its Google Express shopping service, with products available from Whole Foods Market, Costco, Walgreens, PetSmart, and Bed Bath & Beyond at launch, and other retailers added in the following months as new partnerships were formed. The Google Assistant can maintain a shopping list; this was previously done within the note taking service Google Keep, but the feature was moved to Google Express and the Google Home app in April 2017, resulting in a severe loss of functionality.

In May 2017, Google announced that the Assistant would support a keyboard for typed input and visual responses, support identifying objects and gather visual information through the device's camera, and support purchasing products and sending money. Through the use of the keyboard, users can see a history of queries made to the Google Assistant, and edit or delete previous inputs. The Assistant warns against deleting, however, due to its use of previous inputs to generate better answers in the future. In November 2017, it became possible to identify songs currently playing by asking the Assistant.

So next time you are in need of a personnel assistant think again call out to any of the Virtual Assistants already installed on your mobile . A little help from them can go a long way in easing your life. **So sit back relax and enjoy a little Sci-Fi in your real life ... Cheers to the future..**

Moibile Assitant

CHAPTER SEVEN

Digital Twin and Advance Process Controls in Chemical Industry

The world of manufacturing is changing fast and getting competitive. In chemical industry and specifically the specialty chemicals is more true than in any other field. Ask any Specialty chemical manufacturer and he will say they need a system to control highly exothermic reaction within a bandwidth of very minute range while controlling KPI's like temperature, pressure, flow and vacuum. Also, the expectation from management on business out comes like increase in yield, improvement in quality, reduction in utilization of utilities like electricity, compressed gas, water (chilled or brine) etc is very high from the existing control systems installed in the plants.

Once these chemical plants are automated using Distributed Control Systems and PLC to get the next level of optimization and to increase process efficiencies to get focused business outcomes of improved quality and increased yields, they look at implementing Digital Twins and Advance Process Control.

Digital Twin in Chemical Industry

With our experience in process control and digitization we know that process improvements are a long journey with multiple pits stops. This journey can be divided into following stages

1.Plantwide Data Acquisition

This is the very 1st stage of the analytics process. As know you process data is available in multiple formats like process parameter in DCS, PLC. SCADA, textual data or manually entered data, csv files from lab instruments like gas chromatographs and analysers. Some data related to RM

-RM's quantity, initial quality, Order no. batch no. product name, product batch size needs to be taken from SAP. As result data is spread across such data islands. These need to be integrated

2. Historization of Data

The data from all the above-mentioned island needs to be integrated and stored in a single data base. This data being mostly process data needs huge storage space if stored in conventional manner. To store this fast pace data typically in mSec has to be stored over a long period of 2-5 years to make sense when analysing the same. As result most data analytics projects require process historian s with high level of algorithms for data compression and logics like data storage on Rate of Change (RoC) to minimize its footprint. Also process historian is used to retrieve data at high speed for data analysis.

Some of the typical process historians in use are from Osi Pi, Proficy from GE, IP21 from Aspen Tec or Pscope from Microverse Automation.

3. Integration with multiple Automations systems and Islands of information

In any manufacturing set up there are always existing systems be it current updated systems as IEC standards or on legacy systems which are a generation older. But all these systems hold critical data in terms of KPI's like temperature, pressure etc which needs to be collated at one central location. For this integration with all these systems on standard protocols like MODBUS TCP/IP, MODBUS RTU, Profibus, Profinet becomes crucial. The central database for process data has to even integrate with SAP and LAB database to give comprehensive solution.

4. KPI monitoring and dashboarding for actionable insights.

Once data is collated and made available at a centralized database, the next step is to give visibility to the MIS team and top management. This is done by creating KPI dashboards and reports. These dash boards and report can be in standard formats like daily, monthly, yearly or shift wise and using Pi charts and bar chart for better visualization. Beyond this tool like Power BI can be used to give deeper and better meaningful insights to help in quick decision making. Even simple deviation graphs or SPC / SQC graphical dash boards can be made available to the decision makers.

Benefits of KPI monitoring

1. Real time monitoring & control is possible remotely.
2. Improves safety and efficiency of the implemented system
3. Scheduling of production is possible.
4. Risk assessment and muti variate scenarios are possible
5. Customization of services and products is possible.
6. Decision taken using such systems are more informed and efficient in nature
7. Over time prediction of anomalies or deviations in the process is possible.

5. Inferential Data Analytics and offline modelling

The basic level analysis can be done using SPC / SQC reports and KPI dashboards. But if you really want to delve deeper you need tools for inferential data analytics. In this stage historical data is cleaned first to remove unwanted and garbage data and passed through offline modelling techniques. These can be Principal Component Analysis (PCA) or Partial Leases Square (PLS) which models the historical data and gives our 3-4 key KPI's which need minute monitoring to get to the final outcome of increase yields or reduced batch cycle times. There are other techniques like Neural Networks or Fuzzy logic which are advance level analysis modelling to fine to the 1st principle models.

Process modelling Details – Hybrid data analysis and modelling

Step 1 – Big data produced in the process is interpreted by 1st principle physics-based models

Step 2 – The residual data is modelled based on inferential data driven approach.

Step 3 – The non-inferential data which remains is modelled more complex and advance models like Deep Neural Networks (DNN)

Process historian type databases are used for such analytics. They are designed specifically for storing process data and enabled for easy and reliable access to process data by engineers and researchers alike.

Such tools are available in offline software like MATLAB or real time analytics software like CSense Troubleshooter or those from Aspen Tech

6. Corelating data models with Realtime data to get Principal Components or KPI's for deep monitoring

Once the offline models or benchmarks are established , we need to compare Realtime data with these models to get Real time KPI's for deeper monitoring .

7. Online Operator Guidance System for easy decision support.

A Realtime screen can be developed on the SCADA or in Microsoft SSRS to empower the operator or the shift supervisor to take informed decision.

8. Deploying models for Advance Process Control (APC) –

Advance Process Control for process optimization in real time and production scheduling for process efficiency improvement.

RTO – For providing optimized operating condition with satisfying process and economic constraints based on KPI's .

DVR – Data validation & Reconciliation – In this stage data filtering , estimation , smooth and prediction of measures data to ensure the quality measurements in process plant by reducing uncertainties.

Soft Sensors - Process model which uses readily available input variables to predict output variable Same function as physical sensor but values are obtained from a model of the physical sensor

9. RRR – Recheck, rework, redeploy data analytic models for Dynamic process variations.

The advance analytics-based models need RRR methodology to function continuously. The data used needs to rechecked, reworked (analysed), and the models tweaked to co-relate to new dynamics in the process. This new model is now re-deployed to get better and accurate predictions and final outcomes.

The latest next step in the advance analytics and digital transformation is the Digital Twin (DT) . Process industry has always looked at Digital Twin for improving efficiencies and optimize the processes. One of the earliest adopters of computer – based control for safe and efficiency for safe and efficient plant operations. A subset of this is the Digital Twin is the Digital sibling a precursor to this technology and a simple example of which is the Operator Training System (OTS) used to train power plant operators for Realtime condition in plant and how to handle the event before actually running the plant on DCS.

Chemical industry executives acknowledge that data analytics will be crucial for retaining the competitive advantage. With demand for high quality customized products in highly variable batches with short delivery times, the industries are forced to adapt their production and manufacturing style with the current trend of digitalization. Now aim is to achieve operational excellence at operational level with integration of Big Data, IIoT and AI-ML.

Classification based on complexity or maturity of Digital Twin

Step 1 – Pre-Twin – Before physical system is designed. Low fidelity models

Step 2- Digital Twin – Materialized as a set of multiple isolated models

Step 3 – Hybrid Twins – Isolated models are intertwined. Sensor data is Integrated and supervised Artificial Intelligence (AI) is applied to achieve prediction capabilities.

Step 4 – Cognitive Twin or Intelligent Twin – This is an extension of Hybrid Twin with cognitive function. Unsupervised Machine Learning (ML) capabilities are used for self-learning, proactive hybrid systems that optimize its own cognitive abilities.

Artificial Intelligence (AI) in Process Control –

Process Model for Model Predictive Control (MPC) is built using different system identification methods. These models are low fidelity models. The models used within the structure of MPC are accurate, then RTO step can be considered within in the structure of the MPC. The optimization is performed using substitute model instead of the physics-based model. As a result, very less computing is required, even if complex algorithm is required. The newest AI based technique is Deep Reinforcement Learning (DRL) control which provides real-time, self-learning controller in a model free and adaptive environment.

Visualization –

To implement Digital Twin, you need a good front end visualization in terms of KPI dash monitoring & boarding platform. All the components of Digital Twin like Physical part, Virtual part, Connection, Data, and Service have to work perfectly in tandem for a successful DT implementation. The physical world represents the physical assets. The industry 4.0 services are implemented using apps and algorithm. Most of the Digital Transformation service provider like GE Digital Siemens, ANSYS or Oracle have their own digital platform and provide these services on subscription basis or as it is known in the IT world as PaaS, platform as a Service or even in the Software as a Service (SaaS) format. Most of these services are implemented in Oil & Gas, Energy, Metals Mines & Minerals Industry sector. All these digital players have a tie up with Cloud platforms like Microsoft Azure and Google Cloud or Amazon Web Service (AWS). Application is developed and implemented on this cloud platforms. These then act as enablers to deploy Digital Twins.

Conclusion

Overall Digital Twin implementation in the process industry is disjointed across several different topics and industry types. Generic conceptual edifice for Digital Twin in chemical process plant is presented. The digital twin modelling is not a one-time process but a continuously repeating cycle of RRR. These Digital Twin models are Practice loops which are formed to continually control and optimize the chemical process. In future lots of research & development will be done to explore these models along with integration of AI-ML techniques to achieve operational excellence in process plant. Digital Twin implementation is a journey rather than the destination.

CHAPTER EIGHT

DRONES IN AGRICULTURE

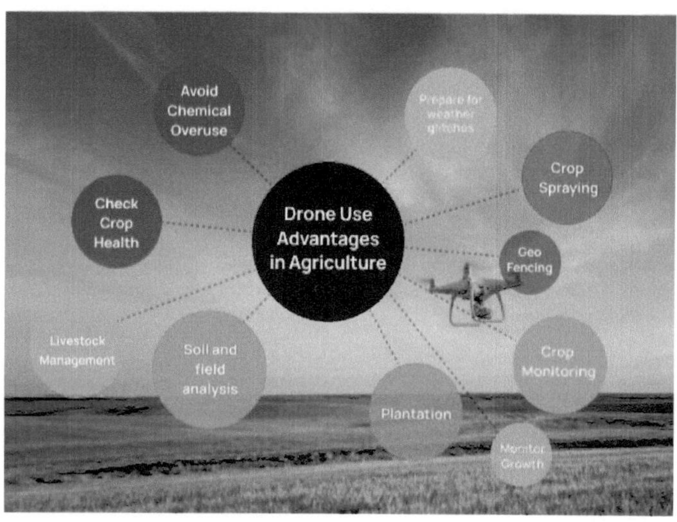

Drones in Agriculture

When you say Drones, the first thing that comes to your mind is a hovering piece of equipment controlled by a remote-control pad or console fitted mostly with cameras used for spy surveillance in War movies. The most recent cinematic reference was one in 3 idiots where the character Ranch fixes the drone being developed by a senior student and taking it for a spin but unfortunately the student who had initially designed and developed the drone does a suicide after getting demoralized and frustrated by his dean's attitude toward his idea. The other was in the successful nationalistic movie URI where the National Security Advisor Mr. Doval chances upon a IIT students concept drone designed for playing and getting away from the boredom he faces in the R&D projects. The drone is then used for surveillance on militants across LOC. This leads to better insights when our Indian Army goes in for a surgical strike.

But drones can do more than this in many sectors where they can be beneficial to the common man. One such us of drones in the agriculture sector. In India which is primarily an agrarian society, such advancement in technology has more use to the farmer than on the borders. For a typical farmer who toils day and night in his fields a technology which can give him a supporting hand in improving his efficiency and yields

is always welcome. In traditional farming there are many pain point , some of the critical ones are

1. increase in input cost due to increased use of chemicals, pesticides and fertilizers.

2. Crop failure due to pests or bad weather conditions

But now with this new age technology of drones there is a helping hand in sight. This can lead to increased yields and better quality of farm produce. Let us explore how drones can benefit an average farmer.

Drones in Agriculture in India

Unmanned Aerial Vehicles (UAV) is the official name of the technology and product but the word drone has been made famous the media. Drones are best used for low altitude arial surveillance. In fact, drones have been in use in Industrial applications in Oil & Gas or in mining sector to monitor high value assets and difficult to reach areas. But its use in Agriculture can benefit the average farmer. Though the drones are new in agriculture there are many start-ups in India who have designed and developed drones specifically to monitor agricultural land and in effective use of crop protection products. This is turn leads to improved productivity and reduction in crop failure.

Use of Agri-drones

There is a lot of focus on use of drones in agriculture by the current government of India . Some of the projects and initiatives taken are

1. On 26th January 2022, the Government of India has also released a certification scheme for agricultural drones, which can now carry a payload that does not include chemicals or other liquids used in spraying drones. Such liquids may be sprayed by following applicable rules and regulations

2. On 23rd January 2022, to promote the use of drones for agricultural purposes and reduce the labour burden on the farmers, the government of India has recently offered, a 100% subsidy or 10 lakhs, whichever is less, up to March 2023 to the Farm Machinery Training and Testing Institutes, ICAR Institutes, Krishi Vigyan Kendra's & State Agriculture Universities.

3. Additionally, a contingency fund of Rs.6000/hectare will also be set up for hiring Drones from Custom Hiring Centres (CHC). The subsidy and the contingency funds will help the farmers access and adopt this extensive

technology at an inexpensive price.

4. On 16th November 2020, the Indian government granted the International Crops Research Institute (ICRISAT), to use of drones for agricultural research activities. With this move, the government hopes to encourage budding researchers and entrepreneurs to look at budget-friendly drone solutions for more than 6.6 lakh Indian villages.

Though the usage will be conditional, yet it is a revolutionary step. Amber Dubey, Joint Secretary, Ministry of Civil Aviation, emphasized that drones are poised to play a big role in agriculture, especially in areas including precision agriculture, improvement in crop yield, and locust control.

Benefits of using drones in agriculture

According to industry experts and reports from international consultants, the global drone market in agriculture is poised to grow at 36% annually and reach USD 6 billion by 2025 itself.

This emerging technology can help reduce time and increase the efficiencies of the farmers. The use of drones in the agricultural sector is only expected to rise as the industry matures, and so it is good to know how to use this technology judiciously.

Some of the best-known uses of drones in agriculture are

1. Soil & field analysis

Soil and field analysis forms an important part of farming. The soil analysis is done to understand the fertility of the soil its mineral content , moisture content , its component mix and texture. All this helps is correct crop selection and also in planning the entire crop management.

The drones in this application are fitted with special sensors to evaluate moisture content in the soil , condition of the overall farm land and many other parameters which help the farmer take correct decision in terms of crop selection and fertilizer and nutrient supplements to be used while farming.

2. Crop monitoring

Monitoring of the crop being farmed becomes an important task. Traditionally this has been a manual activity with the farmer walking across his farm and taking a look at his produce. This again is a person dependant and time-based action. The experienced farmer can spot any anomaly looking at his crop in any stage of growth. But then it is person dependant. With drones and its cameras, the farmer can easily survey his farm in half the time also can detect any signs of insect's infestation or disease very early in the crop life cycle. This results in timely actions and increase in yields. All errors are mitigated which reduces the crop failure. Drones can help in effective crop surveillance by inspecting the field with infrared cameras and based on their real-time information, farmers can take active measures to improve the condition of plants in the field.

3. Planting seeds

Planting seeds or shoots in the farm has always been a time consuming fully manual activity. This can be done effectively in half the time by using drones. The manual labour cost also will go down by using drones. Such drones can replace heavy duty and the costly tractors. These drone can also reduce the carbon footprint since don't use any fossil fuel to run their operation. Drones are more environment friendly.

4. Crop spraying

Traditionally farmers have to carry heavy tanks on their back which are filled with chemicals and fertilizers and then sprayed using hand lever. The farmer has to move from one end of the farm to the other end and then criss-cross across his land. This is a tedious task and very time consuming. Now with drones , their reservoirs can be filled with the correct proportion of chemicals and fertilizers and maneuvered using just a remote control console . an accurate dose of chemicals can be sprayed with a fine mist to effectively cover the entire crop in the farm. The accuracy and precision is very high when drones are used for crop spraying.

5. Check crop health

Farming is a large-scale activity that takes place over acres of land. Constant surveys are necessary to monitor the health of the soil and the crop that has been planted. Manually, this may take days, and even then, there is space for human error. Drones can do the same job in a matter of hours. With infrared mapping, drones can gather information about both the health of the soil and the crop.

6. Decrease overuse of chemicals

Most times farmers don't know how much chemicals to use from crop protection. They go by the education received from the pesticide sales representative or the stockists advice. As a result they may overuse these harmful chemicals with an intension to protect their standing crops but by doing so they are damaging the soil composition and soil nutrients. This is a harmful trend and can affect the crop yields in future . All this can be avoided by using drove technology. Drones can easily detect any signs of pest attacks and can provide accurate data on the degree and range of the pests infestation's. This in turn helps in correct dosage of chemical usage and perfect consistency of chemicals being spayed on the standing crops. As result over use of chemicals is reduced and the harm to crops is minimized.

7. Livestock management

Not only are crops being managed using drones. The technology is now available for even management of livestock. The high resolutions cameras fitted on the droves help in herding together the livestock. It also helps in identifying sick animals which require immediate medical help. They then be given proper medical treatment and also be isolated from the herd so the disease is contained and does not spread to other animals. These drones also help in tracing animals which get separated from the herd and stray away on mountainous terrain. Identification and locating such animals earlier was a huge exercise , with the advent of drones the farm owners can easily locate such animals and bring them back to the farm.

8. Weather anomalies

Weather prediction is very difficult and it the most important factor for any farmer during the entire lifecycle of his crops. Drones can sense the changing environmental conditions and provide data on the weather condition on real time basis. This can help the farmer plant his response and define activities based on these predictions. Sensors are available to check for intensity of the sun rays , wind blowing, rain falling and even the moisture content in the soil . All these can be accurately used to take timely action

9. Growth Monitoring

When everything's going great you don't need minute monitoring or real time monitoring. But things can go wrong in spur of the moment and then it becomes difficult if you don't have a system to monitor your issue. Especially in crop management, surveillance become of prime importance. Even if you need to plan for sales of the crops in the farm market, you need data on the demand and also tune yourself to meet that demand. Using drones you can monitor the crops at all its growth stages, take steps to improve the yields and reduce damages in order to be ready to supply in the market at the right time. The drones can provide precise information to the farmer to take correct decisions. Special sensory cameras can differentiate between healthy and infected crops. For example, stressed crops will reflect less near-infrared light as compared to healthy crops. This difference cannot be detected by the human eye always. But drones can provide this information in the early stages

10. Geofencing

The thermal cameras installed over drones can easily detect animals or human beings. So, drones can guard the fields from external damage caused by animals, especially at night.

Benefits od Agri-Drones

Security

Misuse of drones is minimal as most times they are operated by well-trained drone pilots and as per proper process and policy rules.

Improved efficiency

Using drones, the efficiency of any work on the farm can be doubled with less use of manual labour. As a result, efficiency get increased.

Water Saving

The highest amount of water is used while spraying chemicals and insecticides. But using drones the water content in spraying is highly reduced following the Ultra-Low Volume (ULV) methodology.

Lowest cost of ownership and easy maintenance

With the advent of improved technology and multiple players in its manufacturing the overall cost of the drones has considerably gone down. The maintenance of these new age trends has also become easy due to lower moving parts, cheaper replacement parts and sturdier components being used in the drones. Also, the cost of the activities carried out by drones has become more profitable be it simple surveillance or event the laborious seeding and spraying.

Cons of using drones

Communication and connectivity

Often, online coverage is unavailable in rural areas. Under such circumstances, a farmer needs to invest in internet connectivity, which can turn into a recurring expense.

Weather dependency

Drones are heavily dependent on good weather conditions. Drones don't perform well in extreme conditions of heavy rains or heavy winds. It becomes difficult to fly the drones in such extreme weather conditions.

Skill and Knowledge

Using drones can be highly skill oriented and the knowledge of the technology has to be grasped by the user. An average farmer who is not well educated may struggle to use these drones in the initial phase.

Conclusion

Agricultural sector is never going to be the same with the newer sleeker and sturdier drones being used in most of the farm activities. There are many start-ups in India who are heavily invested in making the agricultural process easy and highly efficient. The technology needs to be made more adaptable to Indian environment and empower the user for better acceptability.

The industry needs to be supported by the government by positive policy frame work and reforms in old and archaic regulations and laws. More needs to be done to optimise the drones to bring large tracts of waste land under farm lands to increase productions and support the ever-growing population with food security.

These drones need, trained pilots and tech savvy newer young age farmers. Which will open up the drone's market. Our farmers and drone enthusiasts are the fulcrum of the new India that our honourable Prime Minister Shri Narendra Modi ji is dreaming about and making that dream a reality is now in our hands.

CHAPTER NINE

Green Hydrogen – The magic fuel of our future

Green Hydrogen

The 2022 will be remembered for the biggest geo political conflict in the world when Russia decided to invade Ukraine and annex certain part bordering Russia. The Russia supreme leader Vladimir Putin has a firm eye on the global gas pipelines and fuel supply chain to Europe fand wants to teach a lesson in war fare and global politics to the Ukrainian President Vlodimir Zelensky's. He wants to break the bravado and adamant attitude of President Zelensky for his vision of joining the NATO.

On the side-lines of this war and geo political crisis, it has become pertinent for the Energy experts across the globe to find an alternative to the fuel crisis that is about to blow in the faces of most countries for their dependency on Ukraine's Gas reserves. More so after a pledge was taken a resolution passed in the UN Climate Change summit to reduce carbon footprint and green house gas emissions by nearly 30% over next 2 decades.

Why Green Hydrogen is perceived to be the cleanest furl for the future

Due to the decreasing cost of renewable fuels like solar, wind and hydro power and the single-minded focus by climate change activist putting pressure on the governments of developed countries to come up with policies and guidelines to curb greenhouse gas emission. This is turn is accelerating potential growth of green hydrogen in political and business perspectives.

Green hydrogen is the latest to be added to the clean energy mix for a sustainable future and cost-effective energy. Green hydrogen is seen as a major component in the decarbonization of energy industry, a clean fuel which will drive the transition and help in shifting from fossil fuel to renewable energy.

Energy transition is in a very early stage, but development of green hydrogen is gaining precedence and gaining acceptance at high speeds. Both the energy markets and projects are gathering pace. Green Hydrogen has potential uses in various end-use sectors, including industry, transport, power and distributed energy.

Green Hydrogen can be produced using multiple processes. One of the most prevalent methods is to split water into hydrogen and oxygen through electrolyser using renewable energy sources. The hydrogen produced using this process is known as Green Hydrogen. There are other methods but they are accompanied by carbon emissions and defeats the entire purpose.

Declining cost of renewable along with the necessity of cubing the green houses emissions is make all of us shift from fossil fuel to this new age fuel. As a result, Green Hydrogen growth is moving upward and making more business sense and is backed by political decision and policies across the globe. Green Hydrogen technology will supply 20 percent of the global energy needs in the near future and will be able to power 450 million vehicles by 2050, this is according to the latest statement by the industry body Hydrogen Council.

The European Energy Commission has put out a roadmap for 2050, they have proposed 85% of energy be produced by renewable, 65% of it will be from Solar and Wind. This will involve a 10X increase in energy storage. To meet this huge demand, the commission has proposed use of excess electricity to split water molecules into hydrogen and oxygen and store hydrogen for using it later.

Most of the Global super powers are aiming to decrease emissions by transitioning entirely to low carbon electricity. Australia aims to generate 100% electricity from low carbon sources by 2030. Fukushima in Japan has proposed similar low carbon energy generation by 2040, Sweden by 2045, California in the US by 2045 and Denmark by 2050.

The reduction in cost of solar and wind-based energy generation is expected to increase their share in the energy basket. Moreover, renewable energy can not only be used to provide low carbon electricity but also generate Green Hydrogen which will finally displace fossil fuel in the end user markets of transport, industries, smart buildings. It can also replace fossil fuels as basic feedstocks in several processes.

This makes green hydrogen a versatile technology that can help decarbonise economies. The cost of electricity is the most important factor in electrolytic hydrogen or Green Hydrogen production. Low carbon hydrogen produced with Carbon Capture Use & Storage (CCUS) or renewable electricity is costlier than hydrogen produced from unabated fossil fuel.

The cost of hydrogen produced from natural gas is generally around USD 2 to 4 / kgH2 , while for Hydrogen generated from renewable electricity (Solar PV or onshore wind) is around USD 2.5 – 6/ kgH2.

Wood Mackenzie, an energy research and consultancy agency has come up with an estimate that Green Hydrogen would be competitive with the gas produced from fossil fuels by 2040. However, Saladvent founder Thierry Leercq a French company experts in hydrogen is very optimistic. During a seminar "Role of Hydrogen in Renewables" conducted by EQ Magazine, he said the combination of super-cheap solar & electrolysis and the ramp-up of midstream infrastructure should decrease the cost of Green Hydrogen to USD 2 per kg by 2025 and then to USD 1 by

2030 matching the cost of natural gas in Europe.

With use of technology and efficient generation of Green Hydrogen using best practices in electrolysis, plant automation, plant load maximization and effects of scaling can reduce the cost to USD 300 /Kilowatt by 2025 at USD 200 /kW by 2030.

The potential scale of Green Hydrogen uses I India is immense and can increase from 3x to 10x by 2050. Facilitating the transition to a carbon – neutral economy, according to the Energy & Resources Institute, a non-profit organization. Nevertheless, several obstacles need to be overcome to tap the full benefits of Green Hydrogen in energy transition.

The barriers foe this technology is less recognition of the importance, the lack of mechanisms to improve & share long term risks of primary large-scale investments, lack of coordinated efforts and actions among all the stake holders, of fair economic treatment of developing technology & limited compliance standards to drive economies of scale.

In the current scenario of climate change and adherence to strict compliance norms, Green Hydrogen has a verry long term advantage and encouraging picture for energy transition. Now that Green Hydrogen has become commercialized in last 3-5 years, the growth potential of this sector is tremendous.

What we are about to see is Green Hydrogen revolution with decrease in cost of generation, improvement in performance and technology in the near future.

The National Green Hydrogen Mission of India has come up with huge numbers for Green Hydrogen production to project India as an export hub & attract global investors as the experts. The Government of India, Union Cabinet approved the mission aimed at decarbonizing the critical industrial sectors like Oil & Gas , Steel , Cement and Chemical companies producing ammonia and methanol. Energy experts when consulted brought to light the great impact this mission will have on Indian economy and Global business environment.

Hydrogen is very volatile and has to be combine with other elements to bring stability. It combines with oxygen to produce water. Electrolyser devices use renewable energy / electricity to split the water into hydrogen and water. The global electrolyser manufacturing capacity has reached only 8 GW till 2021 according to the inter-governmental organization, International Energy Agency. The mission targets manufacturing approximately 8x to 10x this amount to 60-100GW of electrolyser capacity in India. By 2030.

This plan is very ambitious but it is sure attract Global investors, as per energy experts. This is very similar to the 2022 renewable energy target of 175 GW set in 2015 by India. Europe and China have set their targets around 20-25 GW for 2030 in comparison India's target is to reach 60-100 GW which is very high but achievable.

The fundamental feature of the policy is to become a Global leader in production of Green Hydrogen and also an export hub to the Middle East, South East Asia & Africa, If India sets its numbers lower, we will never be able to capture the global hydrogen electrolyser market.

Standard electrolyser manufacturing technologies are already in use, these are alkaline & polymer electrolyte membrane (PEM) electrolysers. Just like China brought in as economy of scale for solar module manufacturing, the focus of the hydrogen policy is to figure out how to reduce the cost is manufacturing & make it mainstream.

The Government of India has allocated INR 19,744 crores toward the mission to develop research & hydrogen hubs, manufacture electrolysers, create job opportunities attract investment and reduce fossil fuel imports. The cabinet approval for the 1st time put clear financials to India's build-out plan for green hydrogen production. The initial outlay will offer good support to launch the industry & will help mature the market. This is similar to how India approached the Solar Power generation market, where the Government backed production linked incentive (PLI) scheme for incentivizing solar panels and module manufacturing in India. Now India is being looked at as an emerging leader in the Green Hydrogen production with clearly defined goals.

Some of the Hydrogen policy's targets are Oil & Gas refineries must replace 30% of fuel usage with green hydrogen by 2035, beginning from 3% in 2025 and huge exponential growth in the production. Fertilizer sector will have target to include 70% green hydrogen by 2035 starting from 15% in 2025. By 2035 urban gas distribution networks should replace 15% their fuel volume with green hydrogen starting from 5% in2025. The proposed capacity of 5 million metric tonnes per annum could potentially replace 30% of the energy produced from liquefied natural gas in terms of the final energy produced from Hydrogen. India consumed 64 billion cubic metric meters of natural gas in 2021-22. In terms of calorific value hydrogen has 2.5 times the energy per tonne compared to natural gas, this means 5 million tonnes of Hydrogen can replace almost 18 billion cubic meters of natural gas. According to energy experts the mission to use Green Hydrogen can be fructified in Oil & Gas refineries and Fertilizers much easily, to achieve the same scenario in steel and cement industry will take much more time.

One of the biggest news came on January 3rd, 2023 the power generation leader in India National Thermal Power Corporation (NTPC) commissioned India's 1st green hydrogen blending project in Gujarat Gas Ltd. (GGL) in Surat. As a follow up on the Green Hydrogen mission, Petroleum & Natural Gas Regulatory Board approved 5% of blending green Hydrogen wit piped natural gas (PNG) initially while adding the blending level may be scaled to reach 20%. This is totally ironic as piped natural gas is a fossil fuel and blending PNG with green Hydrogen is like Green Hydrogen piggybacking on the infrastructure.

Looking at the future demand for Green Hydrogen, India will definitely be a global leader and a bright spot in the global economy.

CHAPTER TEN

Pumped Storage Hydropower – The Newest Energy Source

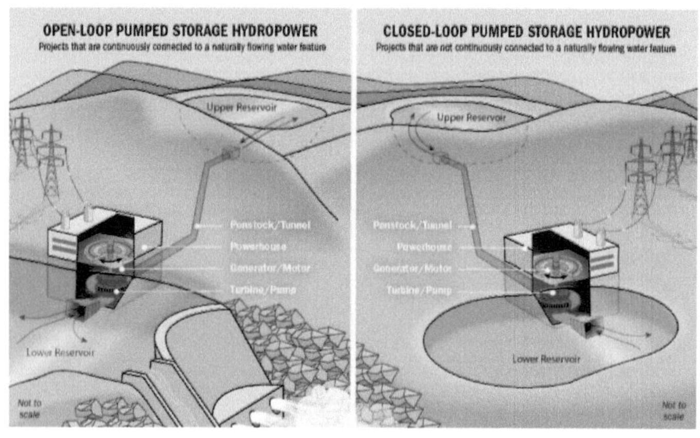

Pumped Storage Hydropower

Day by day as energy consumption grows exponentially both from domestic and industrial users, technologies are getting developed to provide energy using alternative energy resource. Stricter Climate Change resolutions are get passed at the high-powered global committees and at forums like COP26 or even at global economic forums like Davos. The end solution being sustainable energy for all and reduction in cardon emissions with gradual decrease in use of fossil fuel. First it was Renewable energy using Wind and Solar, then energy converted from hot spring of Geo Thermal energy or even the motion of the sea waves has been tried out as alternative fuel – energy. Somewhere in between there was a technology to harness energy from Shale gas in which heavy investment was factored. The latest from Energy technology experts is Green Hydrogen and newest from this stable is Pumped Storage Hydropower.

So what is Pumped Storage Hydropower

Pumped storage hydropower or PSH is a form of hydroelectric energy storage. It is configuration of 2 water reservoirs at different elevations that can generate power a water flows from one reservoir to the second reservoir, the water so flowing is passed through a turbine which in turn generates power. Power is also generated in reverse when the water is pumped from the lower reservoir to the upper reservoir. The overall technology is similar to a giant battery as it can store energy and then power up any equipment at a later time.

Pumped Storage Hydropower was first used way back in 1890 in Italy and Switzerland and later in the US around 1930. Now PSH facilities can be found across the globe. According to the latest Hydropower Market Report published in 2021, PSH accounts for 93% of all utility –

scale energy storage in the USA. America has 43 PSH plants & has the potential to install more PSH plants to double the generation capacity. As per energy reports the US pumped storage hydropower capacity is 22 GW of electricity and 600 GW hour of energy storage facilities in every region of the vast country.

PSH is a clean, flexible and reliable energy and an important player of the electricity grid. PSH has become an extension of the renewable energy sources like wind and solar power and provides energy storage services that integrates all renewable power sources. PSH is a long-range storage commercial technology and an important cog in the renewable power portfolio.

In PSH process water is pumped from the lower water reservoir to the upper water reservoir in the times of high electricity supply, such as during the day, when electricity can be supplied by the Sun's charging of solar panel and /or during low demand. In times of reduced electric supply and /or high demand, such during night times, when some electrical load remains but the sun is not shining and solar energy is inaccessible, water from the upper reservoir is released to the lower reservoir generating electricity as it moves down through a turbine.

How is electricity generated

PSH plants are very similar to traditional hydel power plants, the only difference is that in PSH, we have the ability to generate electricity using the same water over and over again. When power is required to be generated the water flows from the upper reservoir to the lower reservoir due gravity. This water runs the blades of a turbine to generate the required electricity. The water next flows to the lower reservoir and get stored there till the electricity demand get reduced. When the demand is reduced the turbine blades rotate in the reverse direction to pump the water in the upper reservoir so that once again electricity can be generated.

Importance of Pumped Storage Hydropower

Pumped Storage Hydropower (PSH) provides increased capacity, flexibility and balance in energy mix and grid stability. PSH contributes to 95% of storage capacity in the USA. This technology stores energy in the form of water by pumping it to an upper reservoir during times of low demand or high renewable energy output. During high demand or peak time, it generated power by releasing water from the upper reservoir to the lower reservoir through turbine generators.

This technology compliments the current clean energy systems because pumped storage can help integrate the intermittency and seasonality of variable renewables such as wind & solar power., to the grid. This is true for US and the developed nations across the globe which moving towards maximum power generation from clean energy. This is the most economical form of grid energy storage currently available o the US and provides 23 GW of energy and more than 150GW of energy across the world. But there are issues when developing PSH plants there are issues related to licensing and siting to problems with renumeration for the services it can provide. Accordingly, there has been very little new pumped storage development in the US over the past 30 years.

Open loop versus closed loop pumped storage hydropower

The PSH technology is of 2 types open loop or closed loop. Open loop PSH has ongoing hydrologic connection to a natural body of water. With closed loop PSH, reservoir is not connected to an outside body of water.

The water power technologies office (WPTO) invests in new and cutting-edge technology like PSH and research to understand and determine the value of the potential benefits of existing & prospective advanced PSH facilities.

The PSH technology is governed by the International Forum on Pumped Storage Hydropower. This forum focuses on developing guidance and recommendations for PSH to support a transition to a clean energy future. PSH can provide numerous grid benefits, yet it faces many regulatory, economic and siting challenges across the globe.

The International Forum on PSH was founded by the International Hydropower Association (IHA) and the US energy department, the forum established a multi-stakeholder network composed of international government, financial institutions, inter-governmental organizations, and universities along with the industry representatives such as owners, operators, and engineers.

The forum was convened to develop solutions and transfer best practices and experience at the global scale with a mission of delivering sustainable, affordable and reliable power while making contributions towards climate mitigation goals.

Thus, Pumped Storage Hydropower is the newest power source in the clean energy mix and can provide sustainable power to the entire world at very economical cost.

CHAPTER ELEVEN

Optimizing Solar Thermal Power Generation using Advance Analytical Software Systems

Abstract: This paper describes the basic technology behind Solar Thermal Power. How Solar Thermal Power works, the types of Solar thermal collectors and the process of Concentrated Solar Thermal Power (CSP). This paper also gives details of how Power generated from

Concentrated Solar Power is increased using Advance Analytical Software Systems.

Key Word: Solar Radiation, Solar Thermal Power, Concentrated Solar Power, Advance Analytical Software for optimizing CSP.

Introduction: Solar thermal power plants generate electricity indirectly. Heat from the sun's rays is collected and used to heat a fluid. The steam produced from the heated fluid powers a generator that produces electricity. It's similar to the way fossil fuel-burning power plants work except the steam is produced by the collected heat rather than from the combustion of fossil fuels.

Solar Thermal Systems

There are two types of solar thermal systems: passive and active. A passive system requires no equipment, like when heat builds up inside your car when it's left parked in the sun. An active system requires some way to absorb and collect solar radiation and then store it. Solar thermal power plants are active systems, and while there are a few types, there are a few basic similarities: Mirrors reflect and concentrate sunlight, and receivers collect that solar energy and convert it into heat energy.

A generator can then be used to produce electricity from this heat energy. The most common type of solar thermal power plants, including those plants in California's Mojave Desert, use a parabolic trough design to collect the sun's radiation. These collectors are known as linear concentrator systems, and the largest are able to generate 80 megawatts of electricity [source: U.S. Department of Energy. They are shaped like a half-pipe you'd see used for snowboarding or skateboarding, and have linear, parabolic-shaped reflectors covered with more than 900,000 mirrors that are north-south aligned and able to pivot to follow the sun as it moves east to west during the day. Because of its shape, this type of plant can reach operating temperatures of about 750 degrees F (400 degrees C), concentrating the sun's rays at 30 to 100 times their normal intensity onto heat-transfer-fluid or water/steam filled pipes [source: Energy Information Administration. The hot fluid is used to produce steam, and the steam then spins a turbine that powers a generator to make electricity.

While parabolic trough designs can run at full power as solar energy plants, they're more often used as a solar and fossil fuel hybrid, adding fossil fuel capability as backup.

Concentrating Solar Power Technologies

Concentrating solar power (CSP) technologies, sometimes referred to as solar thermal electric technologies, have been developed for power generation applications. Historically, the focus has been on the development of cost-effective solar technologies for large (100 MWe or greater) central power plant applications. The U.S. Department of Energy's (DOE) Solar R&D program focuses on the development of technologies suitable for meeting the power requirements of utilities in the southwestern United States. Numerous solar technologies and variations have been proposed over the last 30 years by industry and researchers in the United States and abroad. The leading CSP candidate technologies for utility-scale applications are parabolic troughs, molten-salt power towers, parabolic dishes with Stirling engines, and concentrating photovoltaics.

Parabolic Troughs

Nine independent power producer (IPP) parabolic trough plants were built during the California renewable energy boom of the late 1980s, and they sell power to SCE. These plants have established an excellent operating track record for this technology. They have delivered power reliably to SCE during the summer on-peak time-of-use period. A number of technology advances have been made in recent years that are expected to make this technology more economically competitive in future projects. Key among these advances is the development of thermal energy storage. A number of new parabolic trough projects are currently in varying stages of project development around the world, some of these will include thermal energy storage

Power Towers

The molten-salt power tower was developed specifically for application in utility-owned solar power stations. These are potentially the most efficient and lowest cost solar power systems. The key feature is the molten-salt working fluid, which provides efficient, low-cost thermal energy storage. This allows solar plants to be designed with high annual capacity factors or used to dispatch power to meet summer and winter peak loads. This technology has not yet been demonstrated in a commercial operating environment. As a result, significant uncertainty exists in the cost and performance of this system. A recent study by Black & Vetch1 classified this technology as being at a pre-commercial status and thus is not yet a candidate for deployment in the commercial power market environment. A number of other power tower configurations are under development. We believe these are either less attractive or less commercially ready than the molten-salt technology.

Parabolic Dishes

Parabolic dishes with Stirling engines are considered attractive because of their modular nature (25-kWe units) and their demonstrated high solar-to-electric efficiency (~30%). Their modular nature means that plants of virtually any size could be built or expanded. These systems do not require water for cooling, which is another benefit in the desert southwest. Unfortunately, the solar application of the Stirling engine was intended to leverage automotive or other applications of this engine, and this in turn would lead to improved engine reliability and reduced cost. The other applications have not occurred to date, and they seem unlikely at present. The Black & Vetch study also found dish technology to be at a pre-commercial status and thus is also not yet a candidate for commercial deployment. Current systems have not demonstrated the level of reliability considered necessary for commercial system.

Concentrating Photovoltaics

Several vendors are currently developing concentrating photovoltaic (CPV) systems. Similar to dish/Stirling systems these systems are considered attractive because of their modular nature (25 to 50kWe units) and their potential for high solar-to-electric efficiency (>30%). These systems also do not require water for cooling. Manufactures are currently providing CPV systems, but only at a few MWe per year and they are still have limited operational experience. Costs are currently somewhere between parabolic trough and flat plate PV. It is our judgment that CPV systems could be attractive for small distributed systems (25kWe and above). It is not clear at what size the economics of a small trough plant becomes the preferred option.

NREL's Recommendation for CSP

Based on the assessment of CSP technologies above, parabolic trough technology is considered the only large-scale (greater than 50 MWe) CSP technology that is available for application in a commercially-financed power project now and in the near future (5 years). The remainder of this report thus focuses on parabolic trough technology.

Parabolic Trough Power Plant Technology

The current state-of-the-art in parabolic trough plant design is an outgrowth of the Luz SEGS power-plant technology. Parabolic trough power plants consist of large fields of parabolic trough collectors, a heat-transfer fluid/ steam generation system, a Rankine steam turbine/ generator cycle, and thermal storage or fossil-fired backup systems (or both). These systems are illustrated schematically in Figure E.3.

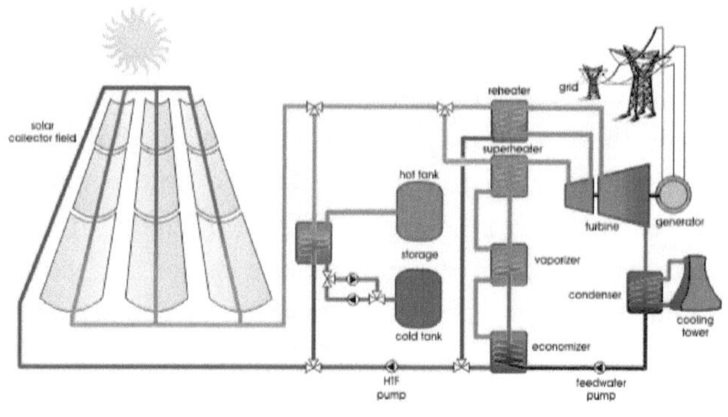

Parabolic Trough Power Plant

Figure E.3: Schematic Flow Diagram of Parabolic Trough Plant

The technology can be described as follows. The solar field is modular in nature, and it comprises many parallel rows of solar collectors aligned on a north-south horizontal axis. The linear parabolic-shaped reflector in each solar collector focuses the sun's direct beam radiation on the linear receiver at the focus of the parabola as seen in Figure E.4. The collectors track the sun from east to west during the day to ensure that the sun is continuously focused on the linear receiver. A heat transfer fluid is heated to 391°C as it circulates through the receiver and returns to a series of heat exchangers in the power block, where the fluid is used to generate high-pressure superheated steam (100 bar, 371°C). The superheated steam is then fed to a conventional reheat steam turbine/generator to produce electricity. The spent steam from the turbine is condensed in a standard condenser and returned to the heat exchangers via condensate and feedwater pumps to be transformed back into steam. Condenser cooling is provided by mechanical draft wet cooling towers. After passing through the HTF side of the solar heat exchangers, the cooled HTF is recirculated through the solar field.

30-MWe SEGS III solar field of parabolic trough

Figure E.5 is a view of the 30-MWe SEGS III solar field of parabolic trough solar collectors at Kramer Junction, California. The figure shows the large field with rows of parabolic trough collectors.

Parabolic Trough Collector Technology

The solar field's basic component is the solar collector assembly (SCA). Each SCA is an independently tracking group of parabolic trough solar collectors made up of parabolic reflectors (mirrors); the metal support structure; the receiver tubes; and the tracking system that includes the drive, sensors, and controls. The solar field in a parabolic trough power plant is made up of hundreds, and potentially thousands, of SCAs

. Table E.2: Luz Solar Collector Assembly (SCA) Characteristics Collector

Table E.2: Luz Solar Collector Assembly (SCA) Characteristics

Collector	Luz LS-1	Luz LS-2	Luz LS-3	EuroTrough ET-100/150	Solargenix DS-1
Year	1984	1988	1989	2004	2004
Area (m_2)	128	235	545	545/817	470
Aperture (m)	2.5	5	5.7	5.7	5
Length (m)	50	48	99	100/150	100
Receiver Diameter (m)	0.042	0.07	0.07	0.07	0.07
Concentratio n Ratio	61:1	71:1	82:1	82:1	71:1
Optical Efficiency	0.734[a]	0.764[a]	0.8[a]	0.78[b]	0.78[b]
Receiver Absorptivity	0.94	0.96	0.96	0.95	0.95
Mirror Reflectivity	0.94	0.94	0.94	0.94	0.94
Receiver Emittance	0.3	0.19	0.19	0.14	0.14
@ Temperature (ºC/ºF)	300/572	350/662	350/662	400/752	400/752
Operating Temp. (ºC/ºF)	307/585	391/735	391/735	391/735	391/735

Luz Table

Table E.2 shows the design characteristics of the three generations of Luz SCAs and two new designs currently under development. The general trend has been to build larger collectors with higher concentration ratios (collector aperture divided by receiver diameter) to maintain the collector thermal efficiency at higher fluid outlet temperatures.

The LS-3 collector was the last design produced by Luz. It was used primarily at the larger 80-MW plants. The LS-3 collector and its components can be described as follows. The LS[1]3 reflectors are made from hot-formed, mirrored glass panels, supported by the truss system that gives the SCA its structural integrity. The aperture or width of the parabolic reflectors is 5.76 m, and the overall SCA length is 95.2 m (net glass). The mirrors are made from a low-iron float glass with a transmissivity of 98%. The mirrors are silvered on the back and then covered with several protective coatings.

The mirrors are heated on accurate parabolic moulds in special ovens to obtain the parabolic shape. Ceramic pads used for mounting the mirrors to the collector structure are attached with a special adhesive. These high-quality mirrors allow 98% of the reflected rays to be incident on the linear receiver. The parabolic trough linear receiver, also referred to as a heat collection element (HCE), is one of the primary reasons for the high efficiency of the Luz parabolic trough collector design. The HCE consists of a 70-mm steel tube with a cermet selective surface, surrounded by an evacuated glass tube.

The HCE incorporates glass-to-metal seals and metal bellows to achieve the vacuum-tight enclosure. The vacuum enclosure serves primarily to protect the selective surface and to reduce heat losses at high operating temperatures. The vacuum in the HCE is maintained at about 0.0001 mm Hg (0.013 Pa). The cermet coating is sputtered onto the steel tube to give it excellent selective heat transfer properties, with an absorptivity of 0.96 for direct beam solar radiation, and a design emissivity of 0.19 at 350ºC (662°F). The outer glass cylinder has an antireflective coating on both surfaces to reduce reflective losses off the glass tube. Getters, metallic substances that are designed to absorb gas molecules, are installed in the vacuum space to absorb hydrogen and other gases that permeate into the vacuum annulus over time. The SCAs rotate around the horizontal north/south axis to track the sun as it moves through the sky during the day. The axis of rotation is at the collector center of mass to minimize the tracking power required.

The drive system uses hydraulic rams to position the collector. A closed-loop tracking system relies on a sun sensor for the precise alignment required to focus the sun on the HCE during operation to within +/- 0.1 degree. The tracking is controlled by a local controller on each SCA. The local controller also monitors the HTF temperature and reports operational status, alarms, and diagnostics to the main solar field control computer in the control room. The SCA is designed for normal operation in winds up to 25 mph (40 km/h) and somewhat reduced accuracy in winds up to 35 mph (56 km/h). The SCAs are designed to withstand a maximum of 70 mph (113 km/h) winds in their stowed position (in which the collector is aimed 30º below the eastern horizon).

Operating Experience of the SEGS Plants

The SEGS plants offer a unique opportunity to examine the operational track record of large parabolic trough plants, Even though the 9 plants in the Mojave Desert of

California with a cumulative capacity of 354 MWe were the first such plants built, they all remain operational (in 2004) and provide an excellent resource for performance and O&M data.

Operations and Maintenance (O&M) of Solar Power Plants

Parabolic trough solar power plants operate similar to other large Rankine steam power plants except that they harvest their thermal energy from a large array of solar collectors. The existing plants operate when the sun shines and shut down or run on fossil backup when the sun is not available. As a result the plants start-up and shutdown on a daily or even more frequent basis. Compared to a base load plant, this introduces additional difficult service requirements for both equipment and O&M crews.

The solar field is operated whenever sufficient direct normal solar radiation is available to collect net positive power. This varies due to weather, time of day, and seasonal effects due to the cosine angle effect on solar collector performance; generally, the lower limit for direct normal radiation in the plane of the collector is about 300 W/m2. Since none of the plants currently have thermal storage7, the power plant must be available and ready to operate when sufficient solar radiation exists. The operators have become very adept at keeping the plant on-line at minimum load through cloud transients to minimize turbine starts, and at starting up the power plant efficiently from cold, warm or hot turbine status.

The O&M of a solar power plant is very similar to other steam power plants that cycle on a daily basis. The plants are staffed with operators 24 hours per day, using a minimal crew at night; and require typical staffing to maintain the power plant and the solar field. Although solar field maintenance requirements are unique in some respects, they utilize many of the same labor crafts as are typically present in conventional steam power plants (e.g., electricians, mechanics, welders). In addition, because the plants are off-line for a portion of each day, operations personnel can help support scheduled and preventive maintenance activities. A unique but straightforward aspect of maintaining solar power plants is the need for periodic cleaning of the solar field mirrors, at a frequency dictated by a tradeoff between performance gain and maintenance cost.

Early SEGS plants suffered from a large number of solar field component failures, power plant equipment not optimized for daily cyclic operation, and operation and maintenance crews inadequately trained for the unique O&M requirements of large solar power plants. Although the later plants and operating experience has resolved many of these issues, the O&M costs at the SEGS plants have been generally higher than Luz expectations. At the Kramer Junction site8, the KJC Operating Company's O&M cost reduction study addressed many of the problems that were causing high O&M costs.

Key accomplishments included:

• Solving HTF pump seal failures resulting from daily thermal and operational cycling of the HTF pumps,

• Reducing HCE failures through improved operational practices and installation procedures Improved mirror wash methods and equipment designed to minimize labor and water requirements and the development of improved reflectivity monitoring tools and procedures that allowed performance based optimization of mirror wash crews, and

• Development of a replacement for flex hoses that uses hard piping and ball joints; resulting in lower replacement costs, improved reliability, and lower pumping parasitics. Another significant focus of the study was the development of improved O&M practices and information systems for better optimization of O&M crews. In this area, important steps were:

- An update of the solar field supervisory control computer located in the control room that controls the collectors in the solar field to improve the functionality of the system for use by operations and maintenance crews,
- The implementation of off-the-shelf power plant computerized maintenance management software to track corrective, preventive, and predictive maintenance for the conventional power plant systems,
- The development of special solar field maintenance management software to handle the unique corrective, preventive, and predictive maintenance requirements of large fields of solar collectors,
- The development of special custom operator reporting software to allow improved tracking and reporting of plant operations and help optimize daily solar and fossil operation of the plants, and
- The development of detailed O&M procedures and training programs for unique solar field equipment and solar operations.

As a result of the KJC Operating Company O&M cost reduction study and other progress made at the SEGS plants, solar plant O&M practices have evolved steadily over the last decade. Cost effectiveness has been improved through better maintenance procedures and approaches, and costs have been reduced at the same time that performance has improved. O&M costs at the SEGS III-VII plants have reduced to about 25 USD/MWh. With larger plants and utilizing many of the lessons learned at the existing plants, expectations are that O&M costs can be reduced to below 10 USD/MWh at future plants

Using advance analytics software like Neural Networks the Solar power generation from CSP power plants can be optimized.

To understand this we need to understand how neural network and Data mining technics work in increasing analysis for process optimization for power generation. In the early planning stage of commercial CSP plants, it is necessary to develop a conceptual plant design that fixes the configuration of the plant, such as the solar field size and the temperature and pressure levels in the water cycle. In the ideal case the design minimizes the levelized cost of electricity (LCOE), describing the costs per generated electricity unit, for the given project and site,and taking specific properties like solar conditions and cooling water availability into account. Simulation-based techno-economical optimization tool for the development of such project[1]specific plant concepts that are well designed with respect to economic criteria. The optimization tool uses adequate Optimization of Concentrating Solar Thermal Power Plants computer science techniques: The global optimum of a solar plant configuration is found using a genetic algorithm, whereas the thermodynamic-energetical procedures of a solar plant are described by an artificial neural network

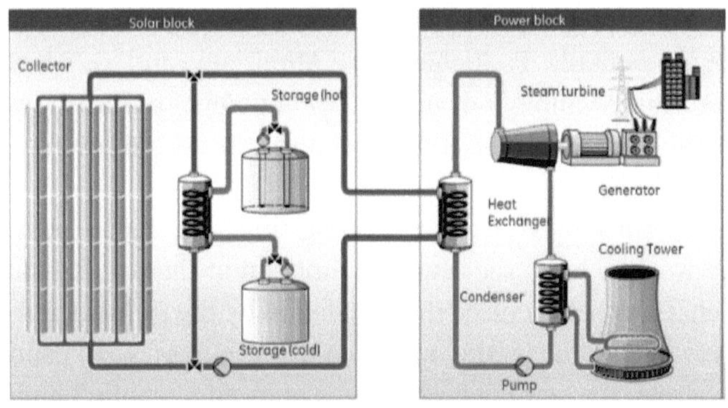

Solar Block Power Block

The physical behaviour of CSP plants is very complex, so their analytical optimization is in practice not possible. Instead we use genetic algorithms for the optimization. The basic idea is very simple: Given an arbitrary set of CSP configurations, we simulate each of them energetically and economically to compute their LCOE average over a year. The LCOE serves as the objective function for the optimization. We select the best configurations with minimal costs and combine them to get a new generation of

configurations for which we repeat the procedure. By mutation this method avoids getting stuck in local minima.

In order to get reasonable LCOE values for a CSP configuration, we need to simulate the plant behaviour for each hour of a year. Without further improvements (see Section 4) the optimization would take about 1000 days due to the time-consuming thermodynamical power block simulation using Thermoflex .

To reduce the calculation time we approximate the thermodynamical power block simulation by bilinear interpolation: For each considered configuration, instead of simulating each hour of a year we simulate only at an experimentally P. Richter, E. ´Abrah´am, and G. Morin determined number of interpolation points. In this way we are able to reduce the running time of the optimization to 2 days.

To further reduce the running time, we train a neural network to learn the required function of the power block behaviour, and replace the simulation by the trained neural network. This is the main contribution of this paper, whereby the computation time is reduced to about 2 hours (including training).

Related work. There are several tools that simulate CSP plants (e.g. Thermoflex). However, these tools are not able to optimise the configurations of CSP plants. Morin [10] connected in his PhD thesis Thermoflex (for power block simulation) with the solar block simulation tool ColSim and with an optimisation algorithm using genetic algorithms and bilinear interpolation (but no neural networks). To our knowledge this is the only work on global optimization of CSP plant designs, including power block design.

There are also papers on combining genetic algorithms with neural networks. The NNGA approach applies neural networks to find advantageous initial sets for the genetic algorithm. In reverse, the so-called GANN approach uses genetic algorithms to set the parameters of the neural network. A broad variety of problems has been investigated by different GANN approaches, such as face recognition, Boolean function learning and robot control classification of the normality of the thyroid gland , color recipe prediction, and many more. In contrast to the above approaches we use in our work neural networks to generate the input data for the genetic algorithms.

The Simulation of CSP Plants

The aim of optimizing concentrating solar thermal power plants is to generate electricity as cheaply as possible. The cost-efficiency of a power plant is generally specified by the so-called levelised cost of electricity (LCOE), which describes the costs per generated electricity unit (e.g. in Eurocent per kWh).

We consider up to 20 design parameters of a CSP plant. Examples for such parameters are solar field size, storage capacity, condenser size, distance between collector rows, as well as pressure and temperature levels. We use pdesign to denote the design parameters of the CSP plant (see Fig. 1 for the CSP plant structure). Our goal is to find a configuration of these parameters that yields a minimal LCOE.

For a fixed configuration of the CSP plant the LCOE must be calculated under consideration of the seasonal and daily variations of its site parameters: The direct normal solar irradiance (DNI) has an influence on the collected thermal power in the solar block and the ambient temperature influences the cooling

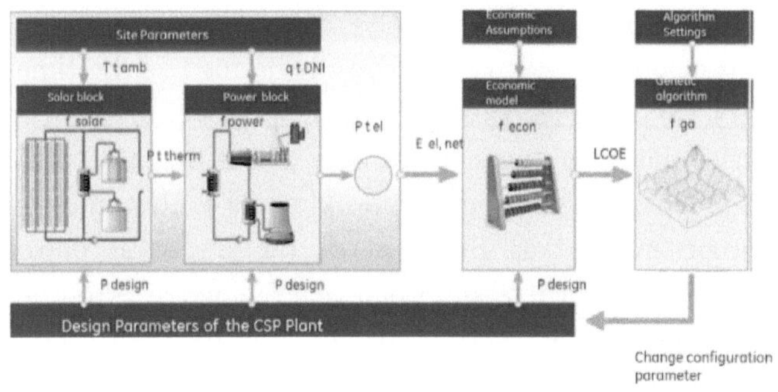

Design Parameter of CSP

Given a configuration fixing the design parameters of the CSP plant, we use the solar and power block models to simulate the CSP behaviour and compute the generated electrical energy for each hour of the year, under consideration of the site parameters. Summing up the generated electrical energy for each hour of the year gives the total energy amount, used as input to the economic model to compute the LCOE under some economic assumptions. The LCOE is the basis for the genetic algorithm to evaluate the configuration of the CSP plant and to create new generations of configurations. Section of the power block1. Therefore, the model for computing the LCOE is based on hourly time resolution over one year. We use qt DNI and T t amb to denote the DNI and the ambient temperature for the tth hour of the year. The LCOE of a CSP plant for a given configuration is computed in two steps:

Firstly, we determine the electrical net energy Eel,net generated over a year using an energetic model of the CSP plant. Secondly, this value is used by the economic plant model to compute the LCOE under consideration of economic assumptions.

The Energetic Model of a CSP Plant

Physically the electrical energy is defined as the time-integral of the electrical power: E = ∫t Pel(t) dt. The electrical net energy generated over a year by a CSP plant for a given configuration is approximated with numerical rectangle method as the sum of the electrical power generated during each hour t of a year: Eel,net = Σ8760

To compute the electrical power Pt el for the tth hour of a year, we first compute with the help of the solar block model the thermal power Pt therm gained by the solar block. This value serves as an input to the power block model that determines the generated power Pt el (see Fig. 2).

The solar block model provides a function fsolar to calculate the thermal power Pt therm = fsolar(T t amb, qt DNI, pdesign) based on the available direct normal solar irradiance qt DNI and the configuration pdesign of the solar block. The calculation first determines the optical collector performance and then subtracts heat loss and thermal inertia effects (when heating up or cooling down).We use the ColSim tool for these calculations. Depending on the operation strategy, either heat is stored or hot fluid is sent to the power block.

The efficiency of converting thermal energy into electrical energy in the power block depends on the thermal power Pt therm, the ambient temperature T t amb (influencing the cooling section), and the configuration pdesign of the plant. The power block model provides a function fpower to specify for each hour t the electrical energy Pt el = fpower(Pt therm, T t amb, pdesign). The power block receives a thermal energy flow from the solar field and/or the storage and converts it first into mechanical and then into electrical energy. Mass and energy balances are computed for each component (e.g., steam turbine, condenser and pumps). We use the Thermoflex tool for these computations.

The total electric net energy Eel,net = _8760 t=1 P t el serves as an input to the economic model.

The Economic Model of a CSP Plant

The economic plant model specifies a function fecon to calculate the LCOE. The total investment costs for the solar block and the power block are computes depending on economic assumptions cecon (e.g. investment costs of collectors,interest rate, etc.) and the configuration pdesign.

The investments occuring at the initial project phase are distributed over the lifetime of a plant using the annuity factor. On top of the investment-related annuity the running costs occuring in the phase of operation of the plant need to be added. The annual running costs consist of: Staff for operation and maintenance of the plant (e.g. mirror washing),water, spare parts and plant insurance. These economic assumptions are included in cecon. The levelised cost of electricity LCOE = fecon(cecon, pdesign, Ptotal el) equals the quotient of the annual costs and the electrical energy generated over a year. The ColSim tool also supports these computations.

Use of Genetic Algorithms to Optimise Solar Plants

As described above the techno-economical model can be used to compute the LCOE of a configuration. However, the number of possible configurations grows exponentially in the number of parameters. To compute the LCOE for every configuration in the search space is not realisable in practice. Hence, we need an efficient heuristic approach to approximate such a multi-dimensional optimisation problem.

Genetic algorithms, a special type of evolutionary algorithms, are well-suited for this purpose because they do not require knowledge about the problem structure. Furthermore, they can easily handle discontinuities, which is of crucial importance here since technically unrealisable configurations have to be sorted out. Another kind of discontinuities comes from several technological solutions (e.g. integer number of collector loops).

A set of initial individuals (in our case configurations) widely spread over the whole search area form an initial generation. Analogous to biological evolution, genetic algorithms select the best individuals (in our case the configurations with the smallest LCOE's) from the current generation and use features such as selection, recombination and mutation to produce a new generation. Iterative application of this procedure leads to individuals close to the optimal solution.

Our optimisation tool embeds such a genetic algorithm. The implementation is based on the free C++ library GALib .We treat undesirable or contradictory configurations by penalisation: they get assigned a very high LCOE and are thus discriminated in the subsequent selection process. On average, the genetic algorithm needs about 25 iterations with 40 configurations per iteration to get close to the global minimum.

Improvements of the Optimisation

The running time of the optimisation based on a genetic algorithm is mainly determined by the simulation of the power block (see Section 2.1) that calculates the generated electrical energy. A typical characteristic diagram for the function fpower of the power block model is shown on Fig. 3. The computation of the function values must consider complex physical processes, and is therefore very time-consuming. For a single configuration, the computation of the electrical energy Pi el generated during a certain hour i of a year needs about 10 seconds2 on a standard computer. That means, it takes about a day to compute the electrical energy Ptotal el generated over a year. The genetic algorithm considers about 1000 configurations until it gets close to the optimum. Thus the optimisation would need around 1000 days without further improvements.

For applications we need to reduce the computation time. Below we present two improvements to speed up the optimisation. The first approach involves bilinear interpolation, whereas the second one employs artificial neural networks.

Bilinear Interpolation

Assume a power block configuration ppower is given. Instead of computing the generated electrical energy Pt el = fpower(Pt therm, T t amb, ppower) for each hour t of a year we compute it only for the grid points of a two-dimensional grid in the space of thermal power and ambient temperature. We use these grid points together with their computed function values to interpolate the function fpower for the given configuration, i.e., to get approximations for Pt el for each hour of the year. We use bilinear interpolation for this purpose, which performs linear interpolation first in one dimension and then in the other dimension. Experiments have shown that it is sufficient to compute the values of fpower for a 4×4 grid, i.e., to simualate a configuration, the modified power block model needs only 16 simulations (10 seconds each) instead of 8760 simulations. As the genetic algorithm needs about 1000 configurations to get close to the optimum, we have an approximate running time of about 2 days

Artificial Neural Networks

To further reduce the computation time, we could think of applying linear interpolation in the dimension of configurations as we did for the environmental state. However, the behaviour in this dimension is highly non-linear, which leads to an unacceptable effect on simulation accuracy.

For this purpose, we use neural networks instead of linear interpolation. Neural networks are able to learn how to approximate functions without knowing the function itself. The learning process is based on a training set consisting of points from the domain of the function as input and corresponding function values as output. We use neural networks to learn the behaviour fpower of the power block model, i.e., to determine the generated electrical energy for a given configuration and for some thermal power and ambient temperature values. There exists a wide range of different neural networks. We use multilayer perceptron, a network consisting of different neuron layers. As shown in Fig. 4, there is a flow of information through a number of hidden layers from the input layer which receives the input, to the output layer which defines the net's output.

Fig. 5 shows the general scheme of the neurons in the layers. Each neuron weights its inputs and combines them to a net input on the basis of a transfer function (usually the sum of the weighted inputs). The activation function determines the activation output under consideration of a threshold value. During the learning process the number of hidden layers and the number of neurons in the layers do not change, but each neuron adopts the weights of its inputs. We train the multilayer perceptron during the optimisation as follows: Before the network is trained we apply bilinear interpolation as before.

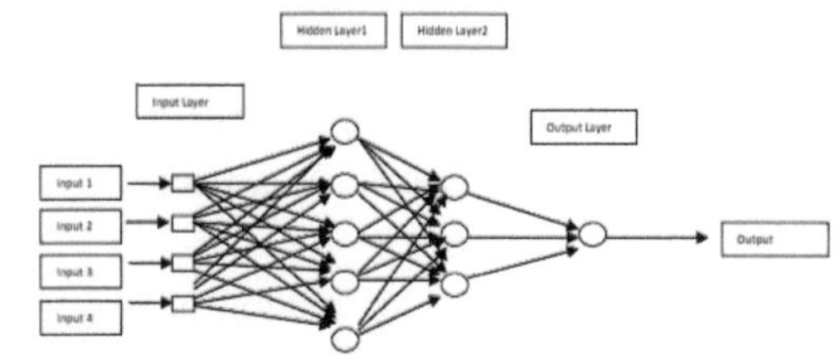

Fig. 4. Topology of an example multilayer neural network with two hidden layers

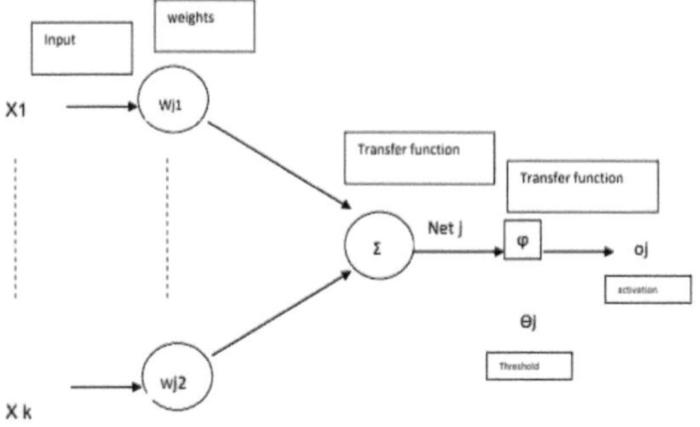

Fig 5 Scheme of an artificial neuron j

Topology and scheme of artificial neuron

The results are used as training set for the network, configuration and environmental state are used as input and the interpolated values as output. Experiments show that it is sufficient to train with the first 20 configurations. After the network has been trained we use the neural network approach instead of the bilinear interpolation.

For each of the 20 configurations we need 16 simulations for the bilinear interpolation, leading to a total simulation time of about 2 hours. The remaining computation times (bilinear interpolation, training, etc.) are insignificant compared to the simulation times. The quality of the results is comparable to the results of the approach using only bilinear approximation.

Experimental Results

We use the Flood tool to define, train, and use multilayer perceptron's with two hidden layers. The networks can be configured by fixing different network parameters. The configuration influences the network's output, which again influences the quality of the optimisation result. To attain good results it is therefore of high importance to find an appropriate configuration. We determined a well-suited network configuration experimentally: For each parameter we trained networks with different parameter values and compared the quality of the outputs. As a measure for the quality of the training we used the relative error between the target and the predicted values for a test set with 12000 data points.

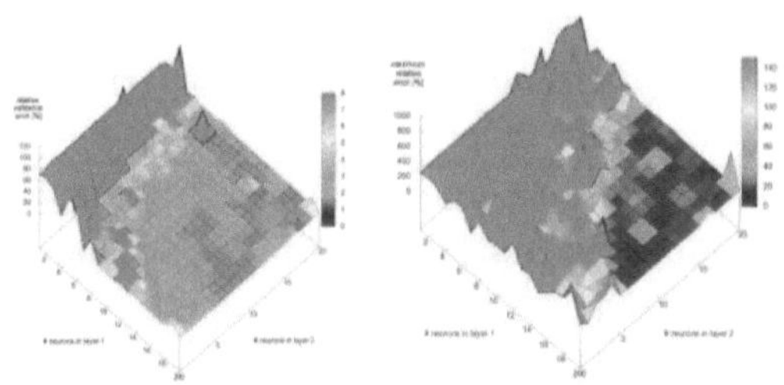

Fig. 6. The relative (left) and maximum relative (right) error for multilayer perceptrons, all having 2 hidden layers but a different number of neurons in the hidden layers

The relative (left) and maximum (right error

During training a network tries to adapt its free parameters (edge weights and neuron thresholds) such that the difference between the target values and the network output for a set of training points gets smaller. There are different training algorithms using different objective error functions. The best predictions were received by application of the conjugate gradient algorithm as training algorithm and the regularised Minkowski error with regularisation weight s = 0.1.

As optimal parameters we determined as optimal parameters the sum of the weighted inputs as the transfer function and the hyperbolic tangent as the activation function for the neurons. We fixed the above parameter values and varied the number of neurons in the hidden layers. Fig. 6 presents the relative validation errors. The best results were gained with 15 neurons in the first and 17 neurons in the second hidden layer with a relative error of 1.2%. With these settings of a neural network, an annual simulation of a CSP plant predicted the LCOE with a relative error of 0.67% and a maximum error of 1.3%.

The effects for the optimisation of using a neural network should be determined in future work. It is expected, that the approximated optimum determined by using a neural network should be close to the optimum found by a time-consuming simulation-based optimisation.

Data Mining using Historians for Analyzing Historical & Real time data:

Typically in a Hybrid power generation set up Solar beams are used to heat special fluid which in turn generate steam and used in power generation.

But sometimes conventional boilers are used in combination with Solar Thermal technology to generate more power.

Continuous and growing increase of fluctuations in electricity consumption brings new hallenges for the control systems of boilers. Conventional power generation will face high demands to ensure the security of energy supply because of increasing share of renewable energy sources like wind and solar power in power production. This can lead to frequent load changes which call for novel control concepts in order to minimize emissions and to sustain high efficiency during load changes.

From combustion point of view the main challenges for the existing boilers are caused by a wider fuel selection (increasing share of low quality fuels), increasing share of bio fuels, and co[1]combustion. In steady operation, combustion is affected by the disturbances in the feed-rate of the fuel and by the incomplete mixing of the fuel in the bed, which may cause changes in the burning rate, oxygen level and increase CO emissions. This is especially important, when considering the new biomass-based fuels, which have increasingly been used to replace coal. These new biofuels are often rather inhomogeneous, which can cause instabilities in the feeding. These fuels are usually also very reactive. Biomass fuels have much higher reactivity compared to coals and the knowledge of the factors affecting the combustion dynamics is important for optimum control. The knowledge of the dynamics of combustion is also important for optimizing load changes.

The development of a set of combined software tools intended for carrying out signal processing tasks with various types of signals has been undertaken at the University of Jyväskylä in cooperation with VTT Processes1. This paper presents the vision of further collaboration aimed to facilitate intelligent analysis of time series data from circulating fluidized bed (CFB) sensors measurements, which would lead to better understanding of underlying processes in the CFB reactor.

Addressing Business Needs with the Data Mining Approach

Currently there are three main topics in CFB combustion technology development; once through steam cycle, scale up (600 – 800 MWe), and oxyfuel combustion. The supercritical CFB combustion utilizes more cleanly, efficiently, and sustainable way coal, biofuels, and multifuel, but need advanced automation and control systems because of their physical peculiarities (relatively small steam volume and absence of a steam drum). Also the fact that fuel, air, and water mass flows are directly proportional to the power output of the boiler sets tight demands for the control system especially in CFB operation where huge amount of solid material exist in the furnace.

When the CFB boilers are becoming larger, not only the mechanical designs but also the understanding of the process and the process conditions affecting heat transfer, flow dynamics, carbon burnout, hydraulic flows etc. have been important factors. Regarding the furnace performance, the larger size increases the horizontal dimensions in the CFB furnace causing concerns on ineffective mixing of combustion air, fuel, and sorbent.

Consequently, new approaches and tools are needed in developing and optimizing the CFB technology considering emissions, combustion process, and furnace scale-up. Fluidization phenomenon is the heart of CFB combustion and for that reason pressure fluctuations in fluidized beds have been widely studied during last decades. Other measurements have not been studied so widely. Underlying the challenging objectives laid down for the CFB boiler development it is important to extract as much as possible information on prevailing process conditions to apply optimization of boiler performance. Instead of individual measurements combination of information from different measurements and their interactions will provide a possibility to deepen the understanding of the process

Conclusion

We described and applied an approach to optimise concentrating solar thermal power plants by determining economically optimal design parameters. The Optimisation of Concentrating Solar Thermal Power Plants combination of simulation, genetic algorithms, bilinear interpolation and neural networks allowed us to reduce the calculation time of the optimisation procedure by around 90% compared to an approach without neural networks. Neural networks were used here to detect complex thermodynamic analogies between different plant designs. The achieved accuracy for prediction of the LCOE is a relative error of less than 1%. We considered simple multilayer perceptron's. In future work here is room for improvements, e.g. using recurrent neural networks, which would need fewer parameters to optimize and due to their feedback, they are likely to suit the problem more

***The above paper was published by Solar Energy Society of India in their ICORE 2012 National Solar Conference Magazine held at Gandhinagar , Gujarat**

CHAPTER TWELVE

Empirical Modelling and Predictive Diagnostics to reduce Operational & Maintenance Cost of Solar Thermal Power Projects

Abstract: This paper describes the basic technology of First Principal Modelling and Empirical Modelling. Power. How Solar Thermal Power works, the types of Solar thermal collectors and the process of Concentrated Solar Thermal Power (CSP). This paper also gives details of how cost of Operational and Maintenance of Concentrated Solar Power Projects is reduced by Predictive Diagnostics Systems.

Key Word: Solar Thermal Power, Concentrated Solar Power, First Principle Modelling, Empirical Modelling, Predictive Diagnostics.

Introduction: Solar thermal power plants generate electricity indirectly. Heat from the sun's rays is collected and used to heat a fluid. The steam produced from the heated fluid powers a generator that produces electricity. It's similar to the way fossil fuel-burning power plants work except the steam is produced by the collected heat rather than from the combustion of fossil fuels.

Solar Thermal Systems

There are two types of solar thermal systems: passive and active. A passive system requires no equipment, like when heat builds up inside your car when it's left parked in the sun. An active system requires some way to absorb and collect solar radiation and then store it.

Solar thermal power plants are active systems, and while there are a few types, there are a few basic similarities: Mirrors reflect and concentrate sunlight, and receivers collect that solar energy and convert it into heat energy. A generator can then be used to produce electricity from this heat energy.

The most common type of solar thermal power plants, including those plants in California's Mojave Desert, use a parabolic trough design to collect the sun's radiation. These collectors are known as linear concentrator systems, and the largest are able to generate 80 megawatts of electricity [source: U.S. Department of Energy. They are shaped like a half-pipe you'd see used for snowboarding or skateboarding, and have linear, parabolic-shaped reflectors covered with more than 900,000 mirrors that are north-south aligned and able to pivot to follow the sun as it moves east to west during the day. Because of its shape, this type of plant can reach operating temperatures of about 750 degrees F (400 degrees C), concentrating the sun's rays at 30 to 100 times their normal intensity onto heat-transfer-fluid or water/steam filled pipes [source: Energy Information Administration. The hot fluid is used to produce steam, and the steam then spins a turbine that powers a generator to make electricity.

While parabolic trough designs can run at full power as solar energy plants, they're more often used as a solar and fossil fuel hybrid, adding fossil fuel capability as backup.

Concentrating Solar Power Technologies

Concentrating solar power (CSP) technologies, sometimes referred to as solar thermal electric technologies, have been developed for power generation applications. Historically, the focus has been on the development of cost-effective solar technologies for large (100 MWe or greater) central power plant applications. The U.S. Department of Energy's (DOE) Solar R&D program focuses on the development of technologies suitable for meeting the power requirements of utilities in the southwestern United States. Numerous solar technologies and variations have been proposed over the last 30 years by industry and researchers in the United States and abroad. The leading CSP candidate technologies for utility-scale applications are parabolic troughs, molten-salt power towers, parabolic dishes with Stirling engines, and concentrating photo voltaic.

Parabolic Troughs

Nine independent power producer (IPP) parabolic trough plants were built during the California renewable energy boom of the late 1980s, and they sell power to SCE. These plants have established an excellent operating track record for this technology. They have delivered power reliably to SCE during the summer on-peak time-of-use period. A number of technology advances have been made in recent years that are expected to make this technology more economically competitive in future projects. Key among these advances is the development of thermal energy storage. A number of new parabolic trough projects are currently in varying stages of project development around the world, some of these will include thermal energy storage

Concentrating Photovoltaics

Several vendors are currently developing concentrating photovoltaic (CPV) systems. Similar to dish/Stirling systems these systems are considered attractive because of their modular nature (25 to 50kWe units) and their potential for high solar-to-electric efficiency (>30%). These systems also do not require water for cooling. Manufactures are currently providing CPV systems, but only at a few MWe per year and they are still have limited operational experience. Costs are currently somewhere between parabolic trough and flat plate PV. It is our judgment that CPV systems could be attractive for small distributed systems (25kWe and above). It is not clear at what size the economics of a small trough plant becomes the preferred option.

NREL's Recommendation for CSP

Based on the assessment of CSP technologies above, parabolic trough technology is considered the only large-scale (greater than 50 MWe) CSP technology that is available for application in a commercially-financed power project now and in the near future (5 years). The remainder of this report thus focuses on parabolic trough technology.

Parabolic Trough Power Plant Technology

The current state-of-the-art in parabolic trough plant design is an outgrowth of the Luz SEGS power-plant technology. Parabolic trough power plants consist of large fields of parabolic trough collectors, a heat-transfer fluid/ steam generation system, a Rankine steam turbine/ generator cycle, and thermal storage or fossil-fired backup systems (or both). These systems are illustrated schematically in Figure E.3.

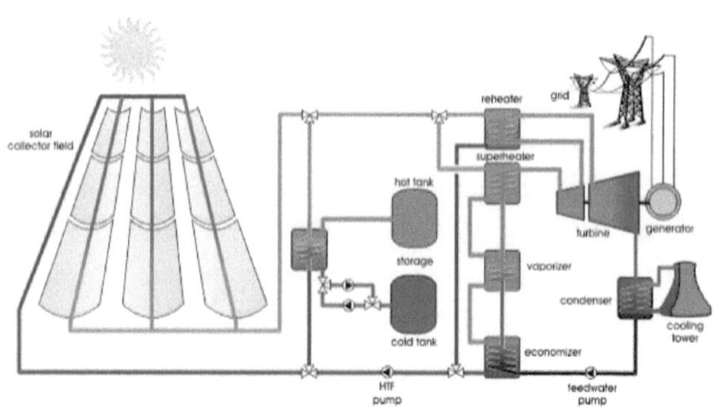

Parabolic Trough Plant

The technology can be described as follows. The solar field is modular in nature, and it comprises many parallel rows of solar collectors aligned on a north-south horizontal axis. The linear parabolic-shaped reflector in each solar collector focuses the sun's direct beam radiation on the linear receiver at the focus of the parabola as seen in Figure E.4. The collectors track the sun from east to west during the day to ensure that the sun is continuously focused on the linear receiver. A heat transfer fluid is heated to 391°C as it circulates through the receiver and returns to a series of heat exchangers in the power block, where the fluid is used to generate high-pressure superheated steam (100 bar, 371°C). The superheated steam is then fed to a conventional reheat steam turbine/generator to produce electricity. The spent steam from the turbine is condensed in a standard condenser and returned to the heat exchangers via condensate and feedwater pumps to be transformed back into steam. Condenser cooling is provided by mechanical draft wet cooling towers. After passing through the HTF side of the solar heat exchangers, the cooled HTF is recirculated through the solar field

30-MWe SEGS III solar field of parabolic trough solar collectors

Heat collection and exchange

More energy is contained in higher frequency light based upon the formula of , where h is the Planck constant and is frequency. Metal collectors down convert higher frequency light by producing a series of Compton shifts into an abundance of lower frequency light. Glass or ceramic coatings with high transmission in the visible and UV and effective absorption in the IR (heat blocking) trap metal absorbed low frequency light from radiation loss. Convection insulation prevents mechanical losses transferred through gas. Once collected as heat, thermos containment efficiency improves significantly with increased size. Unlike Photovoltaic technologies that often degrade under concentrated light, Solar Thermal depends upon light concentration that requires a clear sky to reach suitable temperatures

Heat in a solar thermal system is guided by five basic principles: heat gain; heat transfer; heat storage; heat transport; and heat insulation. [72] Here, heat is the measure of the amount of thermal energy an object contains and is determined by the temperature, mass and specific heat of the object. Solar thermal power plants use heat exchangers that are designed for constant working conditions, to provide heat exchange. Copper heat exchangers are important in solar thermal heating and cooling systems because of copper's high thermal conductivity, resistance to atmospheric and water corrosion, sealing and joining by soldering, and mechanical strength. Copper is used both in receivers and in primary circuits (pipes and heat exchangers for water tanks) of solar thermal water systems.

Heat gain is the heat accumulated from the sun in the system. Solar thermal heat is trapped using the greenhouse effect; the greenhouse effect in this case is the ability of a reflective surface to transmit short wave radiation and reflect long wave radiation. Heat and infrared radiation (IR) are produced when short wave radiation light hits the absorber plate, which is then trapped inside the collector. Fluid, usually water, in the absorber tubes collect the trapped heat and transfer it to a heat storage vault.

Heat is transferred either by conduction or convection. When water is heated, kinetic energy is transferred by conduction to water molecules throughout the medium. These molecules spread their thermal energy by conduction and occupy more space than the cold slow moving molecules above them. The distribution of energy from the rising hot water to the sinking cold water contributes to the convection process. Heat is transferred from the absorber plates of the collector in the fluid by conduction. The collector fluid is circulated through the carrier pipes to the heat transfer vault. Inside the vault, heat is transferred throughout the medium through convection.

Heat storage enables solar thermal plants to produce electricity during hours without sunlight. Heat is transferred to a thermal storage medium in an insulated reservoir during hours with sunlight, and is withdrawn for power generation during hours lacking sunlight. Thermal storage mediums will be discussed in a heat storage section. Rate of heat transfer is related to the conductive and convection medium as well as the temperature differences. Bodies with large temperature differences transfer heat faster than bodies with lower temperature differences.

Heat transport refers to the activity in which heat from a solar collector is transported to the heat storage vault. Heat insulation is vital in both heat transport tubing as well as the storage vault. It prevents heat loss, which in turn relates to energy loss, or decrease in the efficiency of the system

Heat storage to stabilize solar-electric power generation

Heat storage allows a solar thermal plant to produce electricity at night and on overcast days. This allows the use of solar power for base load generation as well as peak power generation, with the potential of displacing both coal- and natural gas-fired power plants. Additionally, the utilization of the generator is higher which reduces cost. Heat is transferred to a thermal storage medium in an insulated reservoir during the day, and withdrawn for power generation at night. Thermal storage media include pressurized steam, concrete, a variety of phase change materials, and molten salts such as calcium, sodium and potassium nitrate.

Operations and Maintenance (O&M) of Solar Power Plants

Parabolic trough solar power plants operate similar to other large Rankine steam power plants except that they harvest their thermal energy from a large array of solar collectors. The existing plants operate when the sun shines and shut down or run on fossil backup when the sun is not available. As a result the plants start-up and shutdown on a daily or even more frequent basis. Compared to a base load plant, this introduces additional difficult service requirements for both equipment and O&M crews. The solar field is operated whenever sufficient direct normal solar radiation is available to collect net positive power. This varies due to weather, time of day, and seasonal effects due to the cosine angle effect on solar collector performance; generally, the lower limit for direct normal radiation in the plane of the collector is about 300 W/m2. Since none of the plants currently have thermal storage7, the power plant must be available and ready to operate when sufficient solar radiation exists. The operators have become very adept at keeping the plant on-line at minimum load through cloud transients to minimize turbine starts, and at starting up the power plant efficiently from cold, warm or hot turbine status.

The O&M of a solar power plant is very similar to other steam power plants that cycle on a daily basis. The plants are staffed with operators 24 hours per day, using a minimal crew at night; and require typical staffing to maintain the power plant and the solar field. Although solar field maintenance requirements are unique in some respects, they utilize many of the same labour crafts as are typically present in conventional steam power plants (e.g., electricians, mechanics, welders). In addition, because the plants are off-line for a portion of each day, operations personnel can help support scheduled and preventive maintenance activities. A unique but straightforward aspect of maintaining solar power plants is the need for periodic cleaning of the solar field mirrors, at a frequency dictated by a trade-off between performance gain and maintenance cost.

Early SEGS plants suffered from a large number of solar field component failures, power plant equipment not optimized for daily cyclic operation, and operation and maintenance crews inadequately trained for the unique O&M requirements of large solar power plants. Although the later plants and operating experience has resolved many of these issues, the O&M costs at the SEGS plants have been generally higher than Luz expectations. At the Kramer Junction site, the KJC Operating Company's O&M cost reduction study addressed many of the problems that were causing high O&M costs.

Key Asset Failures included:

• HTF pump seal failures resulting from daily thermal and operational cycling of the HTF pumps,

• HCE failures due to bad operational practices and installation procedures,

• .Lower Mirror reflectivity due to lack of reflectivity monitoring and improper wash methods followed by mirror wash crew • High replacement costs

· Lower reliability and consistency in the power generation.

· Lower Pumping Parasitics. Another significant focus of the study was the development of improved O&M practices and information systems for better optimization of O&M crews. In this area, important steps were:

• An update of the solar field supervisory control computer located in the control room that controls the collectors in the solar field to improve the functionality of the system for use by operations and maintenance crews,

• The implementation of off-the-shelf power plant computerized maintenance management software to track corrective, preventive, and predictive maintenance for the conventional power plant systems,

• The development of special solar field maintenance management software to handle the unique corrective, preventive, and predictive maintenance requirements of large fields of solar collectors,

• The development of special custom operator reporting software to allow improved tracking and reporting of plant operations and help optimize daily solar and fossil operation of the plants, and

The development of detailed O&M procedures and training programs for unique solar field equipment and solar operations.

As a result of the Operating Company O&M cost reduction study and other progress made at the SEGS plants, solar plant O&M practices have evolved steadily over the last decade. Cost effectiveness has been improved through better maintenance procedures and approaches, and costs have been reduced at the same time that performance has improved. O&M costs at the SEGS III-VII plants have reduced to about 25 USD/MWh. With larger plants and utilizing many of the lessons learned at the existing plants, expectations are that O&M costs can be reduced to below 10 USD/MWh at future plants.

Maintaining critical equipment to ensure high levels of reliability, availability and performance is a primary focus of Power process engineers in every plant today. Without their attention, equipment failure would be rampant. Attention requires frequent, accurate assessment of equipment operating conditions to judge whether equipment meets current production demands and minimizes operational risks of unacceptable schedule interruptions or maintenance costs. In a typical plant, making this assessment involves the collection and effective analysis of reams of data about the health of complex production system elements, like compressors, turbines, pumps and fans.

The amount of data every engineer needs to analyse effectively is growing steadily. Challenging plant economics frequently dictate relentlessly increasing operational demands on ever-aging critical equipment. At the same time, staff budgets are shrinking, experienced people are retiring and new staff need experience to maintain asset health efficiently.

Equipment health and performance data come from periodic and real time systems. Periodic methods for making measurements and analysing particular equipment elements include handheld vibration spectra analysis, oil analysis and thermography and borescope inspection. Such solutions provide a great deal of information about important equipment elements prone to functional failure, but they are time-consuming and intermittent by nature

Reducing O& M costs by Better Monitoring & Diagnostics :

The service architecture of such an O&M operator is summarized in Figure 2

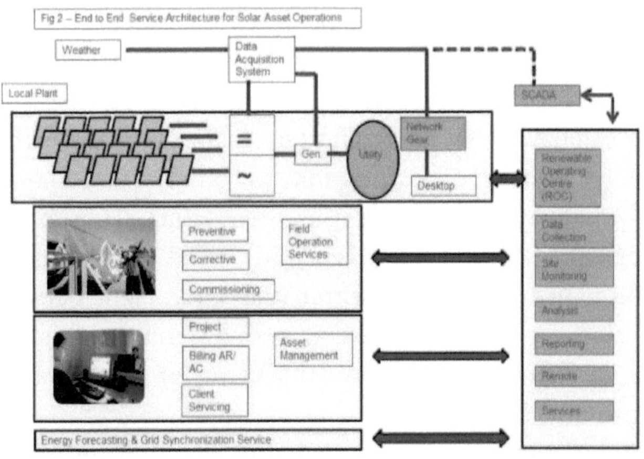

Service Architecture of Solar Power Plant

Reporting:

Periodic reports provide owners/lenders with the necessary visibility in their assets. A detailed and accurate report allows them to account for every unit of energy that the asset is capable of producing. This includes actual energy produced and energy lost because of each failure event along with detailed root cause analysis of the failures. One of the biggest challenges in KPI (PR, CUF etc) reporting for solar assets is proper handling of data gaps. Data gaps could occur on the account of communication failures in the data acquisition system or, sometimes, because of equipment failures (like meter and weather station). There are various ways to handle the data gaps and appropriate analysis is required to either backfill the gaps (using advanced modelling techniques) or ignore the gaps depending upon the extent of the problem. Besides the data gap problem, other uncontrolled parameters (like temperature gradients within the plant, systematic and random errors in measuring instruments, random shading of modules by stray causes) introduce uncertainties in plant performance reporting. A perfectly operating solar asset can have 3-5% uncertainty in single day performance ratio. A longer time duration (a quarter or a year) performance ratio reduces this uncertainty. Hence, this uncertainty should be kept in mind while doing performance assessment within the same plant between two time periods or between two different plants. An accurate report should reflect this error for the benefit of the owners.

Maintenance based on real time monitoring of site conditions (also known as condition-based maintenance):

This is a relatively newer concept in solar CSP assets and allows better efficiencies both in terms of cost reduction and improved energy production. Some of the examples of this approach are modifying the cleaning schedule based on real time soiling losses, scheduling visual checks based on inverter heat sink temperature etc. Such an approach is possible only if the asset is equipped with a robust monitoring and data acquisition system supported by a real time smart alerting engine. For larger sized assets, this is done at the plant location as well as from a remote monitoring centre whereas for smaller size assets (like rooftop installations) remote monitoring centre is almost always the only way to achieve this.

Forecasting:

Solar assets require both short term and long-term forecasting. SLDCs require next day (short term) forecasting at 15-minute intervals from solar assets. As the solar industry scales up to constitute a substantial fraction of the total energy supply, the forecasting will help SLDCs match solar production with other power plants for efficient load dispatching. With the help of an independent weather forecasting service providers an O&M operator can predict energy production based on past behaviour of the solar asset with reasonable accuracy. As performance modelling methods improve short term forecasting accuracy is bound to get better. Long term forecasting is needed by the lenders/owners to correct their financial models for revaluation of the solar asset. A full service operator must be equipped with statistical modelling tools to incorporate system reliability (derived from field failures) into probabilistic production models for accurate long term forecasting.

IT infrastructure and software platform as the backbone of solar assets management

An end-to-end solar asset management requires best in class data management architecture that is smart and scalable. A large-scale solar asset (say a 25 MW plant with string monitoring capabilities) generates approximately 16 GB of time-series data per year. A robust software application should be capable of handling such large amount of data and allow real time troubleshooting and exception handling to detect failures quickly. An O&M operator attends to a failure incident by asking what failed, when it failed, why it failed (root-cause), what is the correction schedule, and what is the level of impact (energy loss and number of equipment failed). Recording all of the above critical information for easy traceability and developing useful business insights through analytics also requires highly customized software tools. In addition, a unified spares management, service ticket management, warranty management applications bring in a complete tool set necessary for an operator to deliver quality service at an optimized cost.

To address the problems that were causing high O & M costs, we need to understand the root cause of asset failures, predict very early such failures and mitigate the issues related to these failures. Technology exists to facilitate prediction of when assets will fail, allowing engineers to target maintenance costs more effectively. Real time systems focused on this area of equipment health monitoring are frequently referred to as equipment condition monitoring (ECM) or predictive asset management (PAM) systems.

This can be done by way of First principle modelling which is a statistical modelling technique to predict failure by way of alarms. The other method and a better way is by empirical modelling, intelligent alarming, predictive diagnostics and alarm prioritization.

First Principle Modelling

This refers to modelling based on First Law of Thermodynamics. First law of thermodynamics: Heat and work are forms of energy transfer. Energy is invariably conserved; however the internal energy of a closed system may change as heat is transferred into or out of the system or work is done on or by the system. It is a convention to say that the work that is done by the system has a positive sign and connotes a transfer of energy from the system to its surroundings, while work done on the system has a negative sign. For example, changes in molecular energy (potential energy), are generally considered to remain within the system. Similarly, the rotational and vibrational energies of polyatomic molecules remain within the system.

From the above, all the energy associated with a system must be accounted for as heat, work, chemical energy etc., thus perpetual motion machines of the first kind, which would do work without using the energy resources of a system, are impossible

Empirical Modelling (EM), is a novel approach to computer-based modelling that developed from research initiated in the early 1980s by Meurig Beynon of the Department of Computer Science at the University of Warwick, England. It has many critics who think of it as a broken type of Functional Programming. Early research within the group led to the development of a new language called Eden - an Evaluator for Definitive Notations. The first implementation of Eden was by Edward Yung in 1987 and a number of contributors have been leading the development of this tool ever since.

The approach of modelling offered by Empirical Modelling (or EM as it is often known) centres on the concepts of Observation, Dependency and Agency. The importance of dependency has been particularly well researched with a number of software tools being developed that exploit dependency maintenance as a native concept.

Most useful among modern condition monitoring methods are empirical regression methods. The empirical methods use historical measurement data for a set of variables known to represent the "inputs" and "outputs" of a system. Parametric methods make some assumptions about the probability distribution from which a data sample is taken – in contrast to nonparametric methods, which do not. The nonparametric model structure is not specified by a functional form, but is determined only by empirical data. Nonparametric methods make few assumptions about the application in question, so they are very versatile. Empirical methods on which commercial condition-monitoring software applications have been developed include parametric methods such as Neural Networks and nonparametric methods such as Principle Components Analysis and Similarity-Based Modelling (SBM).

Key Technology Elements Of A Good Modelling Solution

It was noted above that the choices made in selection of new tools for improving plant performance can be very important to future success of the plant operations. These choices require solid understanding of the problems to be solved and the advantages and trade-offs of potential solutions.

A number of elements to be considered in making selection choices were identified and separated into "technology" elements – judged for suitability to the technical challenge of the application – and "engineering" elements that determine whether a potential solution is a good fit for a particular organization. Here is a brief review of the technology challenges:

Core algorithm accuracy and robustness is a fundamental requirement for analysing complex plant systems. The goal is to provide highly sensitive ability to detect impending problems at their earliest manifestations in the data. Complex plant equipment, like compressors, turbines, pumps and fans, presents wide variation in operating conditions and in the amount of real time instrumentation to indicate operating state, performance and health. The operations environment may strain the instruments' ability to provide high quality data. Any modelling algorithm must, therefore, provide sufficient detection to give early warning of equipment problems despite operational variation, even if some fraction (as much as 25%) of the instruments drift or fail.

Core algorithm execution speed is required for analysing complex systems at high sample rates or when a large number of assets are monitored, such as in monitoring & diagnostics (M&D) centers or decision support centers (DSC). In these environments, instrument counts can be in the tens of thousands, with sample rates as fast as five minutes. Fast execution speeds are characteristic of all of the empirical methods at some scale. But the chosen method must be fast enough for the current monitoring scale requirements, as well as those anticipated in the future. Slower methods are useful only for post-mortem analysis.

Simplicity of model design is a critical requirement for engineers integrating real time monitoring and analytics into their work processes. This requirement ensures that interpretation of analytical results does not require arcane knowledge that is limited to special enclaves in an organization. Models can be designed around ideas about how particular failure modes can be detected and diagnosed.

Simplicity/speed of model training comes into importance during initial model building/implementation, during model retraining following major maintenance and during model rebuilding after equipment overhauls. Identifying the data for an empirical model that best represents full operating range variation must be an intuitive, streamlined process if the methodology is going to fit into an already complex, busy process of operating a plant or DSC . The training process can likely be automated to a large extent; the resultant quality should be easily validated.

Simplicity of model results means that a plant engineer can use the model results directly to reduce the complexity of lots of data for his plant systems and quickly diagnose and prioritize problems - without having to consult experts in statistical or other methods to gain the proper level of understanding. The solution requires the ability not only to detect problems early, but to provide a platform for rapid, certain convergence to the right diagnosis and prognosis.

Visualization/communication of model results is another critical element of the successful methodology applied to complex plant systems. A visual representation can carry a high information density that is easily understood and compactly communicated electronically across an organization. A clear representation of all the data being modelled means that analysis and diagnostics can be accomplished more easily than if the representation is buried in analytical details that have to be massaged out of statistics and then represented in an unfamiliar format

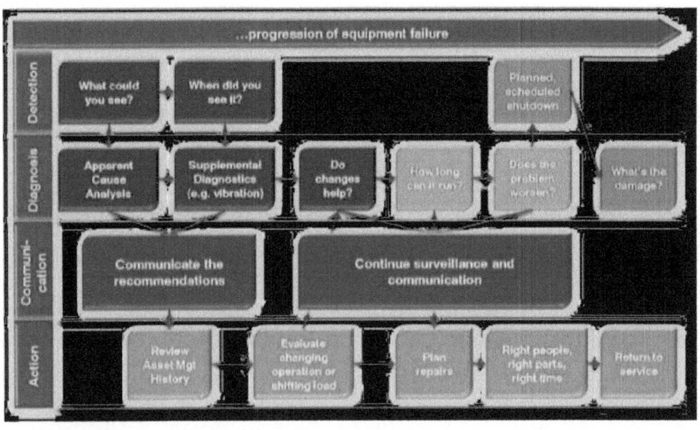

Example of an incident life cycle, showing questions the engineering team asks in the path from problem detection to problem resolution.

Solar Incident Life Cycle

The Best Modelling Solution SBM For Predictive Analytics

Similarity-Based Modelling (SBM) is one of the top technologies used in development of a Predictive Analytics solution to a broad spectrum of real time modelling needs. Other analytical methods failed to satisfy the key technology and engineering requirements outlined above – therefore, they were rejected. Validation of this conclusion may be inherent in the fact that no other analytic method has been applied effectively to ECM on such a broad scale as SBM today. Product development has focused on providing and improving the solution requirements for application in the Power industries. Application in these industries requires demonstration of cost-effective value from the single asset level to a level inclusive of fleets of complex assets across several divisions of global companies.

In these industries, benefits to users include improved understanding of asset readiness, improved maintenance costs and improved resource utilization – all leading to improved business performance in highly competitive industries.

Similarity-Based Modelling (SBM) is a particular form of nonparametric regression. SBM was built as a predictive modelling solution to the need for actionable intelligence from large amounts and diverse sources of current data on equipment like compressors, turbines, pumps and fans brought together in today's complex production systems. SBM models are quickly built from an asset's historical data, with a structure that mimics a natural engineering design. This produces a result that is quickly and efficiently implemented, from a single asset to the largest corporate scope. Using a sample of the data collected from a complex plant system, such as a compressor, a set of "normal" operating conditions can be defined that can be used to reconstruct normal operational behaviour in real time and exclude or flag abnormal behaviour.

The SBM model provides the essential fidelity of the natural system, and it has the advantage of (A) utilizing simple selection guidelines for reference data and (B) requiring no model parameterization. Analytic computation can be done very rapidly, so Predictive Analytics can be applied as effectively to an entire enterprise as to a single asset in a plant. Model design can follow familiar engineering principles, which facilitates interpretation of results and post[1]processing operations, like application of diagnostic logic.

Some of the practical technology benefits of SBM that are not likely to be found in other methods used for equipment monitoring include:

Fast and easy set-up - Fast and easy execution

· Few model design decisions required

· Models based on engineering logic, not arcane statistical concepts

· Simple guidelines for reference data selection

· Computationally expensive modelling processes done off-line and stored

· A clear estimate of normal behaviour

· Works for all equipment, all operating modes

· Easy to interpret results

· Supports automated diagnostics

· Very robust to typical data problems

· Bad data does not disrupt model

· Very tolerant of multiple sensor losses

Brief Review Of SBM for Predictive Analytics

Similarity-Based Modelling (SBM) is a kernel-based, pattern-reconstruction technique using multidimensional interpolation that is designed to exactly fit training data. SBM produces very stable estimates by using a non-linear and nonparametric kernel (similarity operator) to compare new measurements to a set of reference states (state matrix D) – without making stringent requirements on the smoothness and statistical distribution of the data.

The SBM approach measures the input vector's closeness (similarity) to the observation vectors (states) in the D matrix to generate the estimate for that input vector. This has the effect of deriving a current value estimate from the contents of the training space, normalized to the conditions of the current observation. Weight coefficients are computed by solving a system of equations formed using selected reference data points

Several studies have shown that SBM technology outperforms other candidate technologies in detecting faults (1-6). While other nonparametric techniques can produce estimates with similar accuracy metric (measure of how closely the estimates follow the actual), SBM outperforms them in robustness (likelihood that the estimates will over-fit a fault) and "spill over" (the influence a fault in one variable has in the estimates for the other variables)

SBM was designed specifically for the problems of data analysis and diagnostics encountered in real-life equipment. It has been proven through modelling of tens of thousands of assets in power generation, oil & gas, aviation, and transportation applications. It can provide accurate estimates for any number of sensors, of any type, over any load range. Even if a quarter of these sensors fails in the course of operations, SBM still can generate accurate estimates for the remaining sensors without following the faulted signals, making it a very robust methodology. Analytical methods such as PCA and Clustering do not have the same level of accuracy and robustness as modelled signals fail.

Model training is an area in which SBM is strong. Training data can be assembled based on subject-matter expertise from empirical data collected in the plant historian. Because the training data can easily be collected over the range of any independent variables of the system, the effect of these variables can easily be normalized. Automated algorithms can be applied for quickly selecting a set of model conditions that represents the full operating range of the data very well. SBM training is a non-iterative, single-pass operation that involves a single matrix-multiplication and inversion [1]. A model matrix, D, represents the entire dynamic range of the reference behaviour – selected from historical data, personalized to every piece of equipment. An automated SmartSignal[1]proprietary vector selection method is used to build D. The selection algorithm is able to rapidly sort through tens of thousands of observation vectors to construct it. SBM, with the SmartSignal training vector selection algorithm, exhibits a very consistent modelling behaviour. As shown in Figure 4, there is a clear relationship between the model training data and the model (D) structure.

As with the initial model-training process, model retraining is a quick and simple process, taking several minutes to complete. The retraining for a new operating condition involves a simple inclusion of new training vectors into the D matrix from the new operating range via the same vector selection algorithm used to create the

initial model

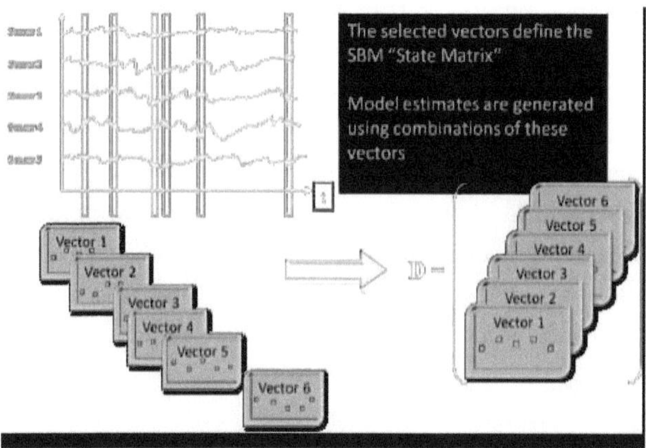

SBM

A distinct advantage of SBM is that model designs normally reflect the most sensible structural elements of the asset being modelled. A design for an SBM model is normally as straightforward as designing a physical model for First Principles methods. The available instruments for an asset can be partitioned into sub-systems that have physical meaning to the engineer. In the example of a steam turbine, the available sensors are grouped such that oil temperatures (OT), metal temperatures (MT) and vibrations (V) are collected to form a mechanical model. Temperatures (T), pressures (P),flows (F) and valve positions (VP) are collected to form individual models of the high-pressure turbine, intermediate pressure turbine and low-pressure turbine. Because SBM was built from the ground up with the detection and analysis of complex plant systems – like compressors, turbines, pumps, fans, and heat exchangers – in mind, it produces results that can be easily interpreted using the subject-matter expertise of the equipment expert, rather than requiring the subject expertise of a statistician or vibration expert.

The simplicity of the SBM model results derives from the simple model structure – this simple structure is not characteristic of statistical methods that have been adapted to real-world modelling problems. Statistical methods such as Clustering-based, Neural-Network-based or Principle Components Analysis-based model systems all can suffer from the problem of design complexity unless this is well handled by the application.

Model design simplicity, the accurate and robust production of estimated normal conditions for each modelled signal and the way this leads to straightforward interpretation of model results, creates another advantage for SBM. The product of the modelling is a set of estimates that mimics the actual data under normal conditions, and it easily shows the trend and magnitude of any differences from normal conditions using a chart format familiar to any plant engineer. This leads to a visualization that complements normal engineering structure, based on familiar failure mode and failure analysis representation. In the example of a tube leak in a heat exchanger, Figure 6, the pattern of the failure easily stands out from the normal condition of the equipment. The overall simplicity of SBM models facilitates development and usage of automated expert rules logic to distinguish between normal operating condition and a faulted condition. It can be used to identify a new operating condition, facilitating automated adaptation of models. Expert rules logic also can be extended to provide fault diagnostics in important cases, like the complex plant systems listed above, where "fingerprints" of failure modes are known from development of subject-matter expertise and knowledge capture

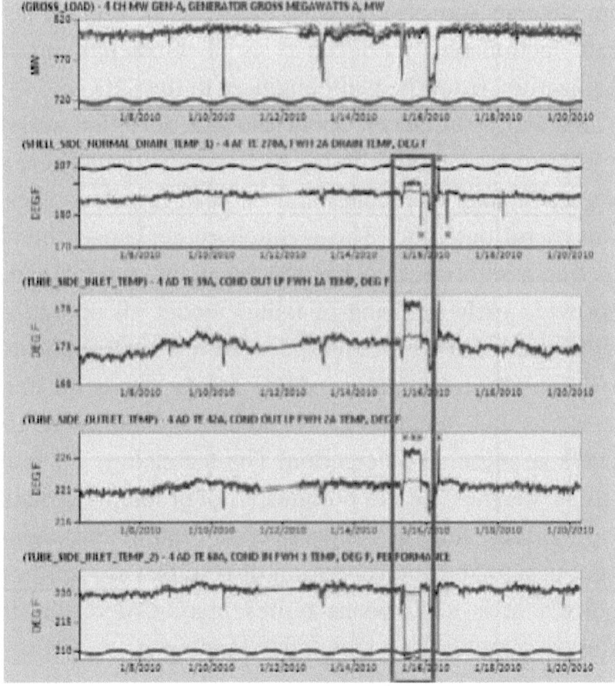

SBM Finger print

CONCLUSION

Engineers are faced with the demanding responsibility of maintaining critical equipment to have high levels of reliability, availability and performance under tight budget constraints. To avoid operating surprises, accurate assessment of equipment operating conditions is needed to judge whether production demands can be satisfied while maintenance costs are controlled. Large volumes of data about the health of complex production system elements are generally available, and the amount of data is growing steadily. Pulling together large amounts of current data from diverse sources across a plant or an enterprise to create actionable intelligence is a challenge to the organization, from the plant engineer to the CIO.

This article has been about the use of digital systems for ECM to create a Predictive Analytic solution. There are both technology challenges and engineering challenges to a successful outcome. The technology challenges involve selecting a solution that has the accuracy and robustness to provide early warning of failure under all operational conditions. It must have the speed for real time application across hundreds or thousands of assets across an entire enterprise. But, technology challenges include more than success at automated detection. The technology must also facilitate diagnostics and prognostics of problems by nature of providing simple connection of equipment design, model design and interpretation of results. One additional requirement is for a visual representation of results that fosters communication and understanding.

In addition to these technology elements, application choice must carefully consider engineering challenges that determine whether a solution will fit normal business practices and thus provide practical results for the business. To satisfy engineering requirements, the solution must be easy to use and have reasonable training demands. This lowers the bar to acceptance when introducing something new to an organization that already is busy. Because there is a wide variety of equipment types and ages in any plant in an enterprise, the solution must be capable of managing this variation, as well as the variation of operating conditions. A single solution that is flexible in its equipment scope and adaptable to the cultures across an organization improves its ROI. It reduces the cost of implementation, maintenance and updating application capabilities. It is critical that any solution chosen must possess the ability to grow with organizational vision – even to help lead it in the case of advanced technology solutions. This requires attention to not only the technology, but also the people and processes supporting its implementation and integration into your organization

Reference – SBM related documents from www.sciencedirect.com

***The above paper was published by Solar Energy Society of India in their ICORE 2013 National Solar Conference Magazine held at KIIT University , Bhubaneshwar , Orissa.**

CHAPTER THIRTEEN

A JOURNEY TOWARDS ACHIEVING OPERATIONAL EXCELLENCE IN CHEMICAL INDUSTRY

Operational Excellence

The Chemical Industry in India is booming especially in speciality chemicals and API's market. With Global supply chains hampered due to the Covid 19 pandemics affecting China and manufacturing getting disturbed due to the on and off lockdowns in Industrial zones in China. The Global manufacturers are looking at China + 1 strategy and India with its long history of chemical manufacturing has come into the lie light.

The current market demands are high and more competitive than in the past. There is a huge pressure on production lines and tough calls need to be taken to improve efficiencies and increase margins. Also, newer start-ups in the specialty chemicals are giving a tough time to the older stalwarts if the game, upping their ante by using technology to bring the right turn around and process driven attitude on the shopfloor. Today the business heads like CEOs of these newer chemical manufacturing organizations are looking at continuous process improvements to reduce the variable cost of productions and increase production simultaneously. Also due to hazardous nature of these chemicals and the exothermic reactions and run-away reactions in the process the use of safety solution become an arear of prime importance.

Opportunities for improvement in Chemical Plants

Every plant operation and production team look at window of opportunity to implement process or operations improvement to achieve their business outcomes. The typical business outcomes and KRA's set by the top management are increase in yields, reduction in input costs, reduction of energy & utilities consumption and improvement in quality of final products

Some of the areas of improvements are listed below

1. Plant production

All products with high demand need have to be produced in higher quantities from the existing assists. For this the process excellence teams are deployed to come up with a process improvement plan. They typically focus is on reducing wastage, getting higher recovery and optimization of existing processes. Most time debottlenecking in the plant production lines can help in getting higher yields from existing assets. In this case the ROI is withing 6 months. Newer technology are availed in terms of advance analytics to get full benefit of improvement in reactions rates and conversions, this can give 2X benefit, one is to reduce variable cost and second is to increase production capacity and better capacity utilization.

2. Raw Material Handling

Raw material (RM) policies and handling is very important for any Chemical manufacturer. Any organization that takes proactive steps to work on RM variability and its solution can be way ahead of their competition in the market. By rule raw materials contributes to 80% – 85% of the total product cost. Thus, reducing the RM cost will automatically have a bearing on the overall cost of production and hence on the bottom line of organization. To increase yields RM variability has to be reduced and also product recovery needs to be improved. For this reactor process parameters need to be monitored minutely, the RM variation needs to be reduced and move towards process optimization. All these will give excellent results to achieve your business outcomes. The parameter variation can be analysed using data analytics and the improvement in consistency can be achieved using golden batch concepts which provides high quality finished products. Some example of process debottlenecking or process improvements is to replace batch reactors with plug flow reactors or loop reactors to get high quality and higher yields.

Reduction in utility consumption

In any chemical process the products manufactured requires huge amount utilities like steam , water , compressed air, chilled water , brine water, fuel oil , furnace oil etc for reactions or end product formation like crystal or powders. These contribute heavily to the cost of production. Among the mentioned utilities the major costs are from steam and power consumption. There is ample of opportunity for optimization in reducing these costs. There are 2 parts to optimizations one is on the generation side the other is on the consumptions side.

On the generation side the major component is the boiler. In boiler there are a lot of opportunities for efficiency improvements by improve the heat exchange and reducing the energy losses. On the plant side or the consumption there are opportunities for effective use of energy using best technology for heat transfer and using waste heat recovery methods. Some of newer technologies used in efficient use of energy are use of heat pumps in distillation and converting low pressure into high pressure steam. Also, use of high-pressure fluid steam to run liquid jet pumps or ejectors can bring in savings in energy consumption.

Typically the power distribution can be optimized by reduction in distribution losses at the transformer end and also improve the transformer efficiencies by minutely monitoring the KPI like transformer oil levels , winding temperature junction temperature etc. At the consumption side the efficiencies of the assets like pumps can be improved by using anti friction lining on the inside casing and impeller of cooling water circulation pumps. There are also opportunities for replacing worn out and legacy motors with high efficiency motors. Analysis of cooling water distribution network will give us insight to reduce the overall power consumption of the plant by minimizing the head loss.

Reducing Effluent Treatment costs

Chemical plants by default have to treat their effluents as per environmental policies and industrial safety policies with which they have to comply. The local pollution control board tracks the effluent discharge and this gets audited regularly. Chemical plant typically has solid, liquid and gaseous discharge that need to be treated before letting it out side the factory. The plant health safety & environment

(HSE) team is responsible for safe and clear discharge of the effluents from the plant complex. This additional cost to make the effluent safe for discharge can be reduced by various methods.

Options can be found to reduce effluent generation at source. For these newer technologies can be adopted like pinch technology to recycle effluent streams within the plant. This reduces fresh water consumption and overall effluent generation quantity. For treatment of effluents, we must find cost effective technologies. The process of effluent treatment is energy intensive so becomes important to develop technology for efficient use energy to reduce the overall cost. The various technologies used in treating chemical effluents are incinerators. Thermal oxidisers, RO plants, MEE and ATFD, catalytic oxidation process.

Other improvement areas

Besides the above-mentioned processes for improvements, there are other areas in the plant where process improvement projects can be taken up by implementing Lean methodology. The different areas are

Supply chain process

The lead time at both the supplier end and the customer end can be reduced. This will reduce the inventory in process and will increase cash flow. Turks and demurrage cost reduction is a great project. Raw material and finished goods freight cost optimization are great opportunities. Since the final products needs to shipped sometimes across continents the packing material used has to be of high quality and need to sustain the journey till end customer location. As a result lower the cost of packaging material also goes a long way in the savings and lower the overall cost of production.

Spare Parts and consumables inventory

Every manufacturing plant has to keep spares and consumables in its inventory to tide over during emergency situations . But keeping this inventory is a cost to the company. There always is an opportunity to keep "Just in Time (JIT) " inventory of spares and consumable to reduce the cost of carrying such an inventory

Finished good & raw material warehouse inventory.

If too much raw material is procured and it lies in the Warehouse or if the products manufactured lie in the warehouse for too log before the are shipped to the end customer , in both the scenarios the cost of maintaining the inventory adds to the cost to the company. So any project which can reduce this cost of carrying additional inventory of raw materials or finished products will add to bottom line of the company.

Workforce productivity

Any manufacturing set up is dependant on its workforce to deliver output. Higher the efficiency of the workforce lower is the cost of production. The workforce efficiency can be studied by doing a time and motion study at the shop floor level. The management can then take decision on how to effectively utilize the man power resource. By following GAMP / GMP guidelines the efficiency of workforce can be substantially increased and this in turn reduces the cost to company.

Process Improvement methodologies

Process improvement at the plant level can be in varied forms. Here we will discuss process improvement practices specially for Chemical plants. The most widely used method is Lean and Six Sigma or DMAIC. These improvement processes can be divided into 2, they can be waste reduction or variance minimization. In case of minimization of waste, Leam manufacturing is applicable, while for variation reduction Six Sigma or DMAIC approach is fruitful. Both methods are frequently used in Chemical Industries.

Lean manufacturing is also better known as Toyota Production System. The Toyota Production System was first implemented in Toyota, Japan in 1930. Taiichi Ohno is the founder of Toyota Production System (TPS) . Six Sigma was invented by Motorola in 1987, this focusses on product quality improvements.

Lean Methodology

Lean is the process improvement technique for continuous waste reduction within a process. Using Lean we can produce products and service faster and at a lower cost. In Lean methodology we can divide various wastes in any process in 7 types abbreviated as DOTWIMP

1.Defects

2. Overproduction

3 Transportation

4. Waiting

5. Inventory

6. Motion

7.Processing

Steps in Lean Methodology

Project Selection

To implement Lean in any manufacturing setup, the first step is selection of the project, we should take up those projects which are of strategic importance for the company. Also, the areas and products where Leam implementation will make a huge difference in commercial terms.

Process Mapping

In this step we need to prepare a process flow / work flow of the existing systems and processes. Each of the process is broken down into steps and smaller activities. We ned to collect data for each of the steps / activities and estimate the cycle time to understand the entire production cycle. This cycle time is the foundation for any improvements to be implemented. Next, we need to segregate these activities in value added and non-value-added categories. Value added activities are those that have commercial implications. There non-value-added activities which are also performed due to regulatory processes like plant safety, environment and health compliance and corporate social responsibility.

Analyse

Once the process flow is defined and ready, the wastages can be identified using the 7-step methodology of DOTWIMP. This gives the manufacturer the opportunity to eliminate the loss due to the wastages, as a result improvements in process take place. This revised process flow is now called the future state value stream map. It is found that as a standard the manufacturing cycle time of the new process is 30% lower than the original process flow map.

Implementation

In the implementation step the solutions for eliminating every type of waste are identified from the previous stages. The master SoP's are revised and the operators are trained on the new SoP's. The improved process is smoothly rolled out on the shop floor which in turn results in improvements in business outcomes of increased yields and reduction in energy consumption 2 of the main objectives which impact the business bottom line. Every process requires a standard change management policy which needs to be followed and regularized so the operators can adapt to the changes and understand their responsibility in achieving the final vision of the organization.

Six Sigma Methodology

Six Sigma by nature is a statistical tool to reduce variance in the process. In this we convert real time problems into statistical problems and then find a statistical solution using statistical tools. This statistical solution is converted to real time solution. Thus Six sigma is a methodology for process improvement based on statistical analysis and implemented through engineered solution.

Lean manufacturing involves minimizing the waste in a process where as in Six Sigma we minimize the variance in the process to improve the product yield or quality and also improve the service response. As per Six Sigma the definition of quality is minimizing variation and getting the same product output every time. Consistency is the keyword in Six Sigma.

Six sigma

The term sigma comes form the Greek alphabet, it is used to denote the data distribution or the data spread over a mean value. Sigma capability is a metric value of the process behaviour. It tells more about how a process behaves in ideal condition and when the process involves variations. Higher the Sigma value, better is the process capability which indicates lower defects and wastages. Sigma is the measure of process variability.

Six Sigma means the process yield are typically 99.99966% or in other word we can say our process has 3.4 defects per million of opportunities.

Six Sigma Steps

Six Sigma is a systematic approach based on statistical tool for reducing process variations. The standard steps in Six Sigma are acronym DMAIC for Define, Measure ,

Analyse, Improve and Conclude The details of each step is explained in brief below

Define

The Define phase is the most important step to start any Six Sigma project. We need to take up projects which are of strategic importance to the management to get best outcomes using Six Sigma process. The project has to be discussed with all the stake holders as this is a team work not an individual project. We must remember that Lean or Six Sigma projects require a cross functional team, because we need people having different skill sets such as operation, statistics, project management, design instrumentation, R&D etc.

Here we identify the project "Y" and collect the data to set the project base line. The final document for Define phase is called Project Charter. A project charter must include below information:

1. **Business case** – why this project is important for the business?

2. **Problem statement** – how big is the problem?

3. **Goal statement** – what is the project including baseline and target?

4. **Project sponsor,** mentor, six sigma black belt person, team member project leader names.

5. **Project time lines** showing start and completion dates for all the DMAIC phases.

6. **Estimated investment** and potential benefits from the project

Measure

In measure phase we identify the X's for our project "Y". In other words we find out independent variable for our dependent variable using various tools such as Fish Bone diagram, Brainstorming, Tree diagram etc.

Model Equation

Once the cause variables are identified, we take up the data collection. Before this MSA is very important to eliminate the error because of measure system. For any measurement process, the instruments and data capturing is an important to estimate the accuracy of the process. After MSA (Measurement System Analysis) , we need to collect the data from the process for independent variables.

Analyse

In the Analyse phase validation is done and verification of all the identified X's using statistical techniques. The cause validating step gives us the effect of X's on the Y. We get a few principal components which are significant and are important for the variation in Y variable. After getting these principal components we do DOE (Design Of Experiments) for our project DOE is the experiment conducted in controlled environment to validate and optimize the independent variables. At the end of the Analyse phase we are with our optimized parameters (X's) which will give the best output (Y)

Improve

In the improve phase, we implement the identified solution after taking inputs form the domain experts. In this phase we need to discuss with the team the solution implementation plan and perform a through risk assessment. After implementation & commissioning we move into the Control phase of the Six sigma methodology.

Control

The changes to the SoP of the process have to be recorded and the plant operators needs to be trained to roll out the process improvements. For this we have to use control charts to see the process variation and compare it before improvement stage. After demonstrating the performance, we close the project and submit complete project report to operation excellence or business excellence head.

Conclusion

Lean manufacturing and Six sigma are 2 of the best practices for process improvement in our plant process. Both these techniques use the power of data analysis. We must learn them as standard practices to implement process improvements on the plant floor and also achieve organizational goals

CHAPTER FOURTEEN

Digital Payments –India Leading the way

Digital Payment - India Leading the way

Just imagine a day in the life of an average middle class consumer shopping for basic things. If you go in the morning to pick up a packet of your daily milk, the shopkeeper asks for Rs. 49 or Rs. 52 per litre depending on which brand you buy. Earlier you had to fish for coins for paying exact change now all you have to do open any of the UPI apps be it Google Pay / PayTM / PhonePe scan the QR code from his sales counter and pay him the exact amount. The money gets transferred from your bank account linked to the mobile number to the shopkeeper's bank account and liked to that QR code. This is so convenient unlike searching for the currency notes and coins to complete the transaction. Same goes for going Dutch with your friends, now that friend who always forgets his wallet when its time to pay the bill at the restaurant has no excuse. You can pay the entire amount and ask your friends to Goole Pay you. The best and most effective way to handle money. But do you know how these digital payments started, what's the technology behind the same. Lets explore in simple language.

The classic definition of digital payments is an electronic transfer of monetary value from one payment account to another using a digital device such as a mobile phone. POS (Point Of Sales) , or computer , a digital channel communications such as a mobile wireless data or SWIFT (Society for the Worldwide Interbank Financial Telecommunication). This also includes payments made with bank transfer, mobile money and payment cards like credit, debit and pre-paid cards.

There is no one definition for digital payments. As in any digital transaction there is an element of partial or complete digital process. For example, a partial digital payment is done when the payer and payee use cash vide third party agents with providers making digital bank transfers in the back end like in case of Western Union. Similarly, a there might be case in which the payer initiates payment digitally to an agent who receives it digitally but the final payee receives the payment in cash from the agent.

So the definition must be fit to purpose. One definition stress the payer -payee interface as the defining element. Another defines digital payments based on the payment instrument, or some other variable. These definitional choices become particularly relevant when the outcome is to estimate the number or share of digital payments in a specific use -case, organization, company, country or region. The definition of digital payments determines how we measure the same.

Benefits of Digital Payments

Whatever is the definition, somethings we are sure about is that Digital payments offers great benefits to its users both individuals and large organizations. Today we have governments or even international development organizations accepting digital payments to complete their transactions.

Following are some of the well-known benefits of Digital payments –

1. Cost Savings – The Digital payments spurs high grade of efficiency and speed which cuts the cost of transaction. A recent report by the Better Than Cash Alliance and the Inter-American Development Bank (ADB) shows the Government in South America for a small country like Peru could save USD 95 million just by shifting all government payments to digital modes available in the market. A huge saving in terms of monetary amounts.

2. Security & Transparency – Due its digital nature any monetary transaction is traceable and accountable. This reduces corruption and theft. For example, a recent financial report analyses risks incurred by the individual purchase officer in diamonds value chain (including theft and assault) due to use of cash. As of March 2020, the Government of India has saved USD 20 billion in social protection payments through Direct Benefit Transfers (DBT).

3. Financial Inclusion – The financial inclusion is due to increase in access to a range of financial services, including savings account, credit and insurance products. The Committee on Payments and Market infrastructure and the World Bank have published a landmark report "Payment Aspects of Financial Inclusion (PAFI). This report outlines how digital payments have helped financial inclusion.

4. **Women's economic participation** - Digital payments have given more control to women over their financial lives and provided them with greater economic independence and larger opportunities. . Digital payments and Digital transactions have contributed to higher women's participation in economic activities than before.

5. **Inclusive Growth** – As per Indian Government's steely resolve for Sabka Saath Sabka Vikas Sabka Vishwas, digital payments in India have helped unlock economic opportunity for the most backward and those who had been financially excluded from the mainstream economy. This in turn has enabled more efficient flow of resources in the economy. There are clear academic evidences that digital payments have led to upliftment of the marginalized persons and reduced overall poverty in India

Digital Payments may not be equal for all

To realize the benefits of Digital payments , the transactions must be done responsibly and in secured ways to protect and promote the safety of he end user. There are guideline available from the Central Banks of countries like Reserve Bank of India , that help define responsible digital payments. The guidelines identify 8 good practices for engaging clients who are sending or receiving digital payments. These also help those who were financially excluded and marginalized due to lack of bank accounts earlier.

These 8 guidelines specifically address the challenges of shifting from cash to digital. There are security and privacy concerns in Digital Payments, these guidelines recommend simple measures to ensure security and improve the confidentiality of user data.

The reliability and resilience of the digital payment system and infrastructure is equally important. System outages might prevent the users from accessing their funds, therefore is imperative that service providers keep the funds safe from cyber crimes and frauds. Robust steps are taken to ensure network reliability and system capacity as well as payments channel secure from fraud , hacking any other form of unauthorized use.

There has also to be transparency in the banking charges and other financial fees to the user while making such digital payments. Lower the transaction cost for both the user and the financial partner higher is the adoption of digital payments and higher is the total transactions which spurs the economy. Affordability and ease of use is the key to sustainability and long-term usage of digital payments. It is also important that these digital payments are designed for gender neutral , so more and more women also start using digital payments and do not get excluded to lack of confidence on technology and its adaptability or access to digital platforms.

Finally the basic mechanism has to be simple, clear and designed is a fashion that its user centric so that every user feels confident to use the digital payment system and also is assured of total security while transacting online.

Digital innovations and technology will continue to improve and grow the payments sector.

The rate at which digital innovation in payment is rising , it has reduced the cost of financial transaction and doubled the compound annual growth rate. It is resulting in new business models and more competition as new players are entering these digital payments market.

Contactless Payments – This is a secure method of payment using a debit or credit or a smartcard enabled by Radiofrequency Identification (RFID) or near field communication (NFC). This type of digital payment is gaining popularity due to speed of transaction and seamless experience.

Open Application Program Interface (API) – A publicly available API provides developers with programming access to a proprietary software applications or web service. Open API's allow new providers to build services on top of the existing platforms. This lowers the barriers of entry for new financial technology players, encouraging competition, innovation and enabling rise of seamless digital payment services for the end-user.

Distributed Ledger Technology (DLT) – DLT is based on the latest block chain technology in which a database is consensually shared and synchronized across multiple sites , financial institutions and even geographies. The database architecture solves the issue of trust among multiple stakeholders and eliminates the "double spend" or double entry problem. This refers to the dilemma of ensuring a digital asset is not spent twice. Since all members of the network hold a copy of the ledger at all times, DLT allows for decentralized digital payment systems that don't rely on one central authority , such as a bank or a public institution.

QR Codes – This is the most prevalent and easy to use during digital payment transaction. A QR code is a 2D Quick Response bar code or a square shaped code that contains data. It ahs become popular as it is simple to use and most easy to exchange information. It has substantially reduced the payment acceptance cost. All that is needed for the payment to take place is a digital device with a camera linked to a bank account.

Biometric payments – Biometric digital payments use Biometric ID as a means of verification and authorization of payment. Biometric ID is any means by which a person is uniquely identified by evaluating one or more biological characteristics like finger print , iris recognition or face recognition or voice recognition.

Central Bank Digital Currencies (CBDC) – Worldwide emerging market economies are moving from conceptual research to intensive practical development of digital currency. Central Banks representing 20% of worlds population say they are likely to issue CBDC in the next few years. Our very own RBI has come up with digital currency **e?-R.** on December 01, 2022. The e?-R is in the form of a digital token that represents legal tender. It is being issued in the same denominations as the paper currency and coins.

India's Central Bank – Reserve Bank of Indi (RBI) spurred a Digital payments revolution

The Reserve Bank of India is Head Quartered in Mumbai since 1981 in a white high rise in the most famous financial district of Mumbai called Fort area, very close the Arabian Sea. The RBI is a pillar for India's rapidly growing digital payments network and a lesson in cooperation between the central bank and private firms.

As per official record India's digital payments volume has increased at a rate of 50% per annum over the last five years. . This the fastest growth rates in comparison to the adoptions in any of the developed countries including USA. The scale of adoption is even greater in India with 1605% annual rate it is the highest globally. This is made possible due to India's own technology platform of unique, mobile enable system the Unified Payments Interface (UPI) the strong backbone of any of the digital payments apps be it Google Pay , PayTM or PhonePe.

Such digital transaction more than doubled to 6 billion in June from the previous year as the number of participating banks jumped 45% to 350. The amounts transacted also doubled in the same period. In addition to this RBI in March introduced a UPI for feature phones, that can potentially connect 400 million users in rural areas.

The UPI system was introduced in 2016. Just after the sudden decision of demonetization Government and RBI followed a policy of withdrawing high denomination currency notes from circulation. UPI was a response to make money transfers easier and safer by allowing multiple bank accounts on the same mobile platform for individual and baseness use alike it rapidly came of age.

UPI was first developed in 2006 when RBI and Indian Banks Association jointly formed the National Payments Corporation Of India (NPCI) . The goal was to be an umbrella institution for digitization of retail payments, and it was incorporated as a non-profit company intended to provide Indian citizens with a easy way to carryout digital transaction. This is an important step for the financial inclusiveness policies the current government is following, a s a result of this Indian economy has grown substantially in comparison to global rate.

Multiple option for payment systems

Growth for individual digital payment users is set to 3X in 5 years to 750 million, according to NPCI CEO Dilip Asbe. Merchant users will double to 100 million. RBI has created an efficient platform and ecosystem for variety of digital payments including RuPay credit and debit cards and now the digital currency e Rupee. They have come with the National Financial Switch cash machine network and payment system using national identity program to bring banking to the door steps of citizens in remote villages

Right from the beginning RBI believed the country like India due to its size and diversity needed multiple payments systems so citizens can choose one from them as per their convenience. UPI is the most innovative and technologically advance system for digital payments backed by RBI and Government of India which wanted to democratize the financial transactions in the hands of the user . UPI today allows the user to make even Rs 1 payment digitally and it is totally free for the consumer in India. The government of India is promoting this payment system with incentives and making more and more merchant to accept digital payments.

With growing cashless society, the traditional way of monetary transaction of using cash is forgotten by country's hundreds of millions of young citizens. It is they who have helped swell the ranks of PayTM and GooglePay to make them world's largest mobile money service providers to more than 500 million users.

Entrepreneurs are using multiple platforms for their financial transactions, one for personal transactions like GooglePay or PayTM and larger gateways like RazorPay or MSwipe for business transaction. With the memory of cash reliance already rapidly fading since the beginning of the smartphone era, the pandemic helped further accelerate the embrace of contactless digital transactions, especially for small amounts, as people tried to protect themselves from the virus.

Open Stack Technology

This transition piggybacked on another unique domestic innovation, the India Stack, a digital identity and payment system built on an open application programming interface, or API. It has been a force for greater financial inclusion by making services easier for consumers to access, including by incorporating the national identity program, Aadhaar, with 1.3 billion users.

Open-stack technology is the foundation of UPI, which transformed India's digital payments, the government's operator of centres for electronic public services in villages and other remote areas. The government decided to promote open-stack technology so that people can try to integrate very quickly and easily. Indian government also promoted private fintech companies, in addition to traditional public sector banks, which allowed quicker adoption of these new age technologies. These services are also available at no cost to the citizen, and that is the uniqueness of India's digital transformation.

Meanwhile, policymakers are planning another big bet on the future of digital money, with even more far-ranging effects on the economy. The RBI is exploring a central bank digital currency (CBDC) designed to meet monetary policy objectives of financial stability and efficient currency and payment operations.

RBI Deputy Governor Rabi Sankar, who oversees payment systems and financial technology, said achieving such an advance would have advantages for currency management, settlement risk, and cross-border payments. He said in a June address at an IMF event on digital money that a digital rupee would have big implications for crypto assets: "CBDCs could actually be able to kill whatever little case there could be for private cryptocurrencies."

Union minister Ashwini Vaishnaw on Friday said India's digital payments transactions last year exceeded those of the combined digital payments of United States, United Kingdom, Germany and France, ANI reported.

"In December 2022, the digital payments transactions amounted to USD 1.5 trillion annualised basis. If you compare total digital transactions in US, UK, Germany and

France and combine them, India's figures are more than that", the minister said at the World Economic Forum summit in Davos.

In recent times, digital form of payments have increased in the country. The most popular forms of digital payments are Bharat Interface for Money-Unified Payments Interface (BHIM-UPI) and Immediate Payments Service (IMPC). UPI has emerged as the preferred mode of payment for users.

The digital payments revolution in India and across the world is here to stay and with UPI it has become more simpler for the common man to now transact anywhere and with no stress over its security from any mobile in his hand. Kudos to NPCI and Government of India to usher in such a technologically advances in a simple app-based solution for everyone to use.

www.ingramcontent.com/pod-product-compliance
Ingram Content Group UK Ltd.
Pitfield, Milton Keynes, MK11 3LW, UK
UKHW040007200726
13854UKWH00001B/83